WORKING2WALK
2012

WORKING2WALK
2012

KATE WILLETTE

Coal Creek Press

Seattle

Copyright © 2013 by Kate Willette

All rights reserved.

ISBN 978-1-304-09495-7

Front cover photograph © Matt Sullivan

Rear cover photograph © Emily Gall

Cover design by Bruce Hanson

Book design by Lynn Kirkpatrick

Published by Coal Creek Press, Seattle Washington USA

and Lulu.com

With thanks and love to my friends at Unite2FightParalysis:

This book is for you.

FANTASTICALLY GENEROUS DONORS

The following names belong to the dozens of people who invested in this project. They believed in its value and they trusted in my ability to make it a reality. I'm so grateful, and I hope the finished product justifies their faith.

David W Hanson	Russell Tuck
Peter Wilderotter	Jim Sampson
Steve Hubbard	Betheny Winkler
Christal Powell	Geoff Kent
Mark Stephan	Pieter de Vries
Martin Codyre	James Shepherd
Alfred R Havel	Paolo Cipolla
Earl Kinsley	Linda Howey
Barbara Carlile	Robert W Mulcahy
Donna Sullivan	Teri Akin
Jerry Silver	Nirav Parikh
Meg Terry	Susan Maus
Randy Neeff	Ross Bentley

GENEROUS DONORS

Sandra Wiercinski	Harvey Sihota
Thomas Borghus	Judy Bentley
Karen Miner	Cheryl Griffin
Susan Hanson	Susan Belauskas
Lynn Kirkpatrick	Corrine Jeanmaire
Carles Alcolea	Jennifer Longdon
Clay Eals	Deborah Davis
Joseph Monteforte	Connie Burris
Margery Patterson	Sue Pendleton
Vanya Sandberg	Kathy Allen
Bob Freel	Christopher Paddon
Cormac McAdam	Ademe Khairakhmetova
Christine Castigliano	Kristin Shamas
Leo Hallan	Belal Breaga Bakht
Jodi Clausing	Cory Maclay
Susan Starbuck	David Zacks
Stephanie Boyles	Matthew Rodreick
Sally King	Joan Griswold
Bruce Cleland	Heidi Venture
Jan Van Pelt	SPACIOUS (Cary Campbell Umhau)
Kabir Kadre	Kate Starling
Arcangela Stefanetti	Barb Schober
Mindy Knadler	Katy Blake
Stephen Feldman	Leah AdangFry
David Duncan	Christopher Paul
Jonathan Washington	Christina Earheart
Ruth Purves	Majed Quaiz

Table of Contents

Introduction 1

Prologue 5
    Working2Walk2012: Where Are We Again? 15

Gene Therapies 23
    Os Steward: Tennis Balls and String 25
    Murray Blackmore: Gene Hunter 38
    SCI Regeneration Mothership: Is It Time? 50

Injury Site Strategies 59
    Damodar Thapa: Paralysis in Nepal 61
    Ravi Bellamkonda: We Need Engineers, Too 67
    Jerry Silver: Patriarch 79
    Injury Site Obstacles: Are We There Yet? 95

Among Friends 103
    Advocacy is the Reason We're Here 105
    Alex Aimetti: Businessmen in Our Corner 116
    Leif Havton: Cauda Equina Injuries 123
    Up Close: Os, Jerry, and Murray 131

Bench to Bedside 143
    Jonathan Thomas: Follow the Money 145
    Aileen Anderson: Start with Good Science 153
    Stephen Huhn: Human Trials for Chronics 164
    Considering Combinations 172

Bridge Builders **181**
    Jerry Silver: Bypass the Whole Damn Thing 183
    Justin Brown: Low-hanging Fruit 195
    We Just Gotta Get Some Axons to Go 203

More Stem Cells **213**
    Mark Tuszynski: A Living Relay System 214
    Hans Keirstead: Something Old, Something New 225
    Shots on Goal 237

Please Come to Boston **243**

Sources and Resources **247**

Acknowledgments **257**

Introduction

A couple of months ago I was talking to a friend who works with spinal-cord-injured people in a local rehab facility. I cringed when she told me she had a patient who was planning a trip to another country for stem cell treatment, but I said only, "I hope it's safe." I was thinking of another conversation I'd had with a different friend. This one had taken his paralyzed son to a different foreign country, also for a treatment not available in the USA. That man had talked about the importance of hope and the need to feel that he was doing *something*, whatever the risk or the cost. He talked about the camaraderie and the strange, brave innocence that's part of seeing yourself as a pioneer. You'll know things other people don't. You'll be like Magellan, whose crew first made it all the way around globe in 1522. You might even recover some function.

Going overseas for treatment is something my own husband once considered, so it's not like I don't understand what drives people to take these kinds of chances.

We look at the reality of living out life with bodies so damaged, and it makes a sort of sense. At the very least, it would be oh-so-satisfying to prove the inscrutable doctors wrong; it's only human to want to try. One purpose of this book is to lay out the long version of what I would like to say to people who might be thinking of making such a decision. The science around spinal cord injury is challenging, even for nerds like myself – *but it's not impossible to understand.* I got my own ticket punched for this ride in March of 2001, when a grim ER physician told me bluntly that my young husband was never going to feel or move anything below the level of his nipples again. The doctor was wrong, as it happened, but the ride was nevertheless on. That night, my knowledge of biology ended with whatever they taught in my 1968 high school class, not that I remembered a word of it.

The other reason I decided to write this book is that I live for stories. I wanted to tell the amazing story of Working2Walk, and I wanted to do it in a format that was roomy enough to capture both context and meaning. I've reported via live blog about the conference – i.e., in real time – for the last several years. I've read a lot of good summaries from other attendees after the fact, and I've published a few articles of my own. The peer-reviewed papers that bring scientists to the attention of Working2Walk organizers are written for other researchers working in other labs; they're not really for us. The youtube videos of Working2Walk scientists presenting their data and taking questions from participants give viewers a broad sense of what direction the research is going and of what sort of men and women are engaged in the long effort toward a cure. Taken together, all these resources present a strong picture of what this event is about. *The problem is that they rarely are taken together;* they're far more likely to be absorbed piecemeal and over time. At the conference itself, the sheer mass of information can knock you flat.

Sooner or later, there will be effective therapies to people who live with paralysis. I'd like it to be sooner, right, and I'm convinced that well-informed advocates have a major part to play in the process. The scientists all say so, every chance they get. They need us, but we won't be able to be very helpful if we can't tell a stem cell from a tomato. My position is that if we can get a basic understanding of what they're doing and why, we'll be much better advocates on our own behalf.

I also just want to say this: there's a real person on the other side of these very imperfect pages. I tried to write as if I were speaking to the woman I was that night in the ER, somebody who was so clueless that she didn't even know what she didn't know.

I tried to talk to her in an ordinary voice, making sure as I went that she wasn't getting lost in an onslaught of information. That's what this book is: a guide that covers the basics, as seen through the lens of a single two-day event. Thanks for reading.

Prologue

> *Spinal cord injury is a rough way to meet people,*
> *but I've met some keepers!*
> Betheny Winkler

Fall, 2004 at CareCureCommunity. "CC," as we all referred to it, was an early example of the promise of social media. Before twitter, youtube, instagram, tumblr, or facebook, and long before the proliferation of political and personal blogs, CC was an online message board. It was a set of forums that invited all comers to choose a name, pick an avatar, and join the conversation. Everybody there had one thing in common: we were all either paralyzed ourselves or very close to someone who was. The website was a project of Dr. Wise Young, who led the W.M. Keck Center for Collaborative Neuroscience at Rutgers University. To this day, Wise himself is often online there, answering questions, telling stories, and offering encouragement.

There was always plenty to discuss – from the puzzle of how to use a catheter on an airplane (just put a blanket over your lap and hope your neighbors don't get the wrong idea) to how much you should do for someone who was both disabled and lazy (as little as possible) to the latest political storm (George W. Bush's stem cell policies were much on our minds). You could discuss movies, or trade recipes, or share insights about managing pain. There was a sort of magic in the way that people's personalities came through the screen, so that you would often forget completely that the person making you laugh or cringe or cry was typing with a pencil stuck between their teeth, or attached to a useless hand.

Some personalities stood out, like my friend Betheny Winkler. Funny, sharp, warm, and sometimes unbearably real, she was the neighbor you were glad to see, the sister whose phone calls lifted your day, the sardonically helpful voice in your ear when things got brutal. I once spent an evening talking with her about the weird ways that spinal cord injuries manifest themselves. My husband could not wriggle the toes on his right foot, but he was able to take steps unassisted. Betheny had full use of both her hands. She could stand up and even stagger around a bit, but she couldn't close her eyes in the shower to wash her hair. Her proprioception – the ability to know where your body parts are without looking – was gone, and closing her eyes made her dizzy. Both her patriarchal dad and her 100% redneck husband, she told me, were impatient with her need for the wheelchair, which made her sad; she'd tried so hard not to depend on it. I had no idea what she looked like and knew I'd probably never meet her, but she was a friend like no other at a time when I badly needed someone to make me laugh at the stupid and endless wrong that is spinal cord injury.

* * *

It would be hard to overstate how hard we at CC took the death of Christopher Reeve that October. He had been the public face of our hope – the embodiment of the possibility of making paralyzed bodies work again. Most people who sustain serious damage to their spinal cords have a story about the moment when doctors flatly informed them that they would never walk again. It's a brutal moment, but at CC the prevailing opinion was that a cure was within reach if only enough resources were directed toward achieving it. According to Wise, it would be the result of a series of breakthroughs, and it would arrive as a complicated array of complementary therapies. Most of all, we were aware that achieving those breakthroughs and testing the dosage and timing of those therapies would take money and a network of human trials.

> Christopher Reeve. Where were we when he needed us to back him up? He was too much a lone voice in the wilderness. I feel great shame that I didn't step up to the plate and help when he was alive. Now is the time to make amends. For those who think he was only in it for himself, I have this to say: he could have chosen, like so many do, to sit back and continually throw himself a big pity party. He did not.
>
> Susan Maus, January 2005

When Reeve died, the foundation that bears his name was in the process of trying to pass legislation (the Christopher Reeve Paralysis Act) to set up the conditions for creating just those conditions. A network of clinical trials and the money to fund them: that was the goal.

That's why when a little group of people at CC began to think about a rally in Washington DC to honor Reeve, the focus quickly became legislative. People wanted to meet each other and see the faces that went with the screen names, but they knew there needed to be some kind of agenda beyond that if they really were going to pick up where Reeve had left off. The thread that launched what became Working2Walk was posted on January 12, 2005.

(The author of that post, known as bigbob, wasn't known for brevity,

> CareCure Forums > SCI Community Forums > Cure
> **Couldn't we (CareCure members) organize a Cure SCI Washington March?**

but he did have bluster.) Betheny was one of the many people who quickly took up the idea, gleefully imagining the possibility of becoming a burr under the saddle of then-Senator Sam Brownback, who had solemnly informed the world that spinal cord injury could be cured without embryonic stem cells. (He knew this for sure.)

Another woman who would become important in the push for advocacy was my friend Susan Maus. Injured in 2000, she was the mother of two young daughters and part of a warm, hard-working, Minnesota-nice extended family. Her high cervical injury kept her from feeding herself, scratching her own nose, doing any of her own personal care, or shrugging her shoulders. It didn't keep her from holding down a job, mothering her kids, being a wife to her husband, or taking up the cause of the rally. In her no-bullshit, straightforward, starchy way, she was the very first to reply to bigbob's question:

We'd need to develop a purpose statement to rally around – IMO, full passage of the Christopher Reeve Paralysis Act (including funding) and establishment of a US clinical trial network to test therapies here that are currently sitting on shelves collecting dust. Then, we'll need to clearly and rationally prepare a cost/benefit analysis (including cost to society for care and treatment of SCI in dollars). With our budget pressures as they are, new funding is not very realistic. We'll need to demonstrate why funds should be diverted from another study or program. Presenting the information en masse in Washington would give us great momentum to push forward. Uniting with others such as CRPF and SCI Society would be beneficial as well.

And that – remarkably – is how it started. Someone asked a question, and someone else saw what the next steps would have to be. Hundreds of people around the country nodded at their screens. This could work.

Not, however, without some drama and trauma. These were, remember, people who had never met one another; what they shared was access to the internet and the bum luck to be paralyzed. The fact that they were online talking about a crazy plan to get a bunch of para and quadriplegics to the US Capitol to demand action meant that they were also self-selected to be both passionate and vocal. What's more, because spinal cord injury respects none of the usual social boundaries, they were diverse in terms of politics, religion, ethnicity, education, gender, class, and age. If they'd been all in one room doing the initial planning, there would still have been disagreements but at least there would have been body language and eye contact to help keep things civil. Adding to the spectacle was the ruling ethos at CC, that conversations should be frank and open, with as little interference from moderators as possible.

There were very public fights over basic questions: Who had the most reliable data about how many people were living with injury? Who could say with authority what it cost the nation to care for those unable to work? What exactly should be the request of congressmembers? Should the rally include some kind of stunt that would be likely to attract attention from the press, ala the famous ACT-UP scenes on behalf of AIDS? Was it acceptable to bring politics into the frame? What other groups would make good allies? Most difficult of all, should the rally focus on embryonic stem cell research or leave that contentious issue off the table? There was no one in charge to make those calls except the people who assumed authority themselves and were able to make their cases most convincingly to those reading the arguments as they unfolded.

This is where Betheny's brash, happy-warrior optimism mattered. Her sense of humor would pop off the screen, defusing tension and refocusing people on the fact that this plan was ridiculous, impossible, and actually happening. It was happening. She'd refer to herself as a pesky crip, and joke that people ought to carry signs to congress: *Which one of you boneheads tabled the Christopher Reeve Paralysis Act?* Exactly thirty days after bigbob's original question, they had a name: Cure Paralysis Now. There was a rally website, a fixed date, and dozens of wheelers signed up. People were volunteering to do everything from printing flyers to researching hotel accessibility to finding volunteer assistants for those who would need them. They were raising money.

In counterpoint to Betheny, Susan was a determinedly rational, down-to-earth, conservative voice. Joining the two of them in leadership were four more CC regulars, all female. Pam, known as Princess Leia at CC, was partially paralyzed by a spinal tumor in 1999. She worked her ass off on that rally, according to those who watched it unfold. Suzanne was the Chinese mother of a paralyzed son whose determination to make a contribution to curing paralysis

> Cure Paralysis Now proudly announces its first Washington Rally, to be held April 12, 2005 in our nation's capital. CPN is a grassroots organization . . . we are proud to announce that Christopher Reeve's widow, Dana, has agreed to join us and speak at the Rally.
>
> Marilyn Smith

brought her all over the planet. Faye was the single mother of a paralyzed son and two other children – Dutch-born, self-made, and furious over stem cell policies in the USA, she was a born fighter, for better or for worse.

The sixth member of the steering committee was a woman who preferred to keep her head down and stay out of the online bickering. Marilyn Smith didn't take up a lot of emotional space or get involved in feuds; she just quietly did what needed doing and then asked for the next job. Her ticket to the paralysis ride got punched second-hand, on the day her son, Noah, was injured in a freak car crash on his way back to college after a visit home for the Thanksgiving holidays. A tire flew off the axle of an oncoming truck. It bounced across the median and flew directly into the windshield of Noah's car, and so at twenty he became one of the people facing a doctor and hearing that he was sentenced to living in a marginally functional body. Marilyn and her husband, John, were midwesterners who had settled in a small Oregon town; they'd spent their adult lives living deliberately and raising Noah and his younger brother, Isaac. Noah's injury pulled his mom – who had always been willing to raise her hand when work needed doing – into a whole new world of activism.

Before Reeve's death, she could be found at CC trying to get her head around the state of activism aimed at a cure. Two weeks after bigbob posted his idea, with a date set, a speaker list started, a dozen ideas competing for the agenda, and hotel search underway, she was volunteering to handle donations and suggesting that it was time to form a steering committee. The plane had clearly lifted off, but there was no landing gear, or even a blueprint for it. Marilyn was exactly the right person to sharpen the drawing pencils and clear a space for the thankless grunt work of chasing details. These six women – two of them paralyzed, four of them mothers of paralyzed sons – were the original organizing committee of what would become Unite2FightParalysis.

* * *

Against all odds, the rally itself was an unqualified success. Friendships were forged and the idea of an empowered community took hold. The original organizers came to Washington DC with the hope that they could collectively pick up where Reeve had left off – by learning to advocate for a cure and helping to pass the legislation that bore his name. They left with a plan. Next year they'd gather again – this time not just to rally, but to become educated. They would hold a science and advocacy symposium, where scientists would be invited to help people learn about the latest research. They decided to call the event Working2Walk.

chapterOne

Working2Walk2012: Where Are We Again?

All things happen in a particular time and place, and understanding *where* and *when* you are is an important part of every story. The 2012 conference was at the Hilton Hotel that sits just across the highway from the Orange County John Wayne Airport in Irvine, California. Nearby to the north is the Disneyland complex in Anaheim, and to the west the famously upscale beach communities of Huntington and Newport. If you like kids, flying into John Wayne is fun; your plane is certain to have an unusually large number of small passengers, all of whom will be in a good mood. Southern California weather in early November is typically fine, and that would be the case most of the time at Working2Walk.

Unfortunately things were very different on the east coast; Hurricane Sandy made landfall in Atlantic City, New Jersey just hours before travelers began arriving for the conference. Sandy was an eight hundred-mile-wide storm that destroyed whole neighborhoods and knocked out power for weeks for millions of Americans. Several people who had registered to come to Irvine found themselves trapped on the east coast while the airlines scrambled to get their flight schedules back in order.

The other major news event that week was the general election in the USA, which would be held three days after Working2Walk ended. It's safe to say that no matter which outcomes attendees were hoping for, *everyone* was looking forward to seeing the endless campaign season come to a conclusion at last. We were thus sandwiched tightly in *time* between a storm and an election, and loosely in *place* between a spectacular surfing beach and America's most famous theme park.

Why Irvine, California? Since the first rally back in 2005, Working2Walk had been held once every year, almost always east of the Mississippi. At first that choice was driven by a legislative agenda that centered around the Christopher Reeve Paralysis Act; there was a pressing need for conference attendees to lobby members of congress in person. On March 30, 2009, newly-elected President Barack Obama signed an omnibus bill that contained the act as one of its titles, thus ending the years-long fight to see it become law and freeing Working2Walk organizers to take the conference on the road. Unite2FightParalysis board member Todd Phillips – a young New Yorker whose only connection to the paralysis community was a passionate admiration for Christopher Reeve – had spent much of the previous summer calling every member of Congress to ask that they sponsor the bill.

With no congressional agenda to push, Working2Walk began to travel, always looking for locations that combined accessibility, proximity to major research institutions, and the possibility of combining forces with established local advocates. Irvine had all three, in spades.

It's a modern city, at least compared to Washington DC, with most development having taken place after WWII; as recently as the 1930s most of the land that became the city of Irvine was still ranch and farm country. This matters only because there are fewer buildings and public spaces that aren't governed by the rules that guarantee accessibility, which have been enforced on new construction projects since the Americans with Disabilities Act became law in 1990. Entrances to buildings are flat, or there are smooth, slow-rising ramps to make getting in and out safe. There are curb cuts at every intersection.

The John Wayne airport itself is small and perfect. It has lots of flights, it's easy to navigate, and it's built to accommodate people with young children in strollers – all of which means that it works for wheelers as well. Anyone who's ever had to navigate a fifteen-minute walk through a major airport at the end of a long trip can appreciate how good it feels to see the exit doors as you come out of your gate. There are a number of major corporations headquartered near that airport, and that guarantees there will also be hotels with meeting spaces.

Choosing a particular hotel can be an exercise in patience, though, involving careful explanations to event planning staff about the physical requirements. The fortunate majority of people haven't had to think through what it means to travel in a wheelchair, especially without full use of their hands. For some, the word, "accessible" simply means that there are plenty of grab bars in the bathroom. For many conference attendees, though, that word defines a very specific set of constraints. It means that all the doors open with push bars and the lighting, phone, and alarm controls are within easy reach. It means there's room to park a wheelchair next to the bed.

In the bathroom, it means that there's no bathtub, no lip into the shower, and room under the sink for the front of the chair and your knees. It means a tiled, roll-in shower, or a sturdy shower chair that can be placed so that the water actually reaches you, and faucet handles that can be used without working fingers. It means the racks inside the closet are low enough to reach, and the flooring material is not soft, thick carpet, which makes pushing a wheelchair like walking through quicksand. The height of the bed is important to those who can't lift themselves to transfer into and out of it; it has to be as close as possible to the height of the wheelchair. There have to be enough rooms that fit the basic requirements, but there never are. Usually that means some of the participants need to stay at a nearby hotel and struggle with getting into and out of a van in order to be at the meetings.

All this is to say that it's understandable why most people living with paralysis don't travel often. It's complicated. Difficult. There has to be a very good reason . . . like a conference where you can meet and question the scientists working on a cure. And that was Irvine's most important qualification to host Working2Walk – the collection of neurobiologists whose labs were housed just five miles away, on the campus of the University of California, Irvine.

*   *   *

Here's a true story. In 1878, an immigrant sheep rancher named James Irvine completed a series of transactions that made him the sole owner of 110,000 acres, stretching twenty-three miles from the Santa Ana River to the Pacific Ocean—the land that would become Orange County, California. His great-granddaughter, Joan grew up on this land; she was the sort of young girl who loved horses, and she spent her childhood roaming the property from high in a saddle. In 1996, the grown-up version of that child happened to see a well-known paralyzed actor doing a television interview.

His disability was unimaginable. He couldn't be alone, ever, because he depended on a machine to push air in and out of his lungs, and if the machine failed he wouldn't have the breath to call for help. Below his neck, his body was not moving. At all. Not taking orders from his obviously agile mind, and not sending signals of pain or pleasure to that mind. The interviewer questioned him about the moment of his injury, which had happened the previous year when his beloved horse balked at a jump. The man won the woman viewer's admiration because he declined to take the interviewer's suggestion that what had happened was the horse's fault. He spoke instead about his hopes for a cure.

The woman reached out to the man, offering to donate $1 million as seed money for a research institute that would be located on the campus of the California university that was named for her family, on the condition that he join her in raising funds to build the institute. When Christopher Reeve's wife, Dana, received this offer by mail, she assumed it was from a crazy person and set it aside. Two months later, Joan Irvine Smith tried again, and this time she reached

Reeve. They dedicated the site for the new Reeve-Irvine Research Center nine months later.

* * *

In a way, the original intent of those enthusiastic CC members, planning their little rally to honor Reeve's courage and commit to living up to his example, was about to be fully realized. They had managed to gather enough strength in numbers to be, if not his voice, then his ears and eyes. And they were going to be joined by a number of California scientists whose careers had thrived because Christopher Reeve wanted to get well so badly. The Reeve-Irvine Research Institute opened its doors in 1999, having hired Dr. Os Steward as its director. And that Dr. Steward is about to give the opening science presentation at the 2012 Working2Walk.

Gene Therapies

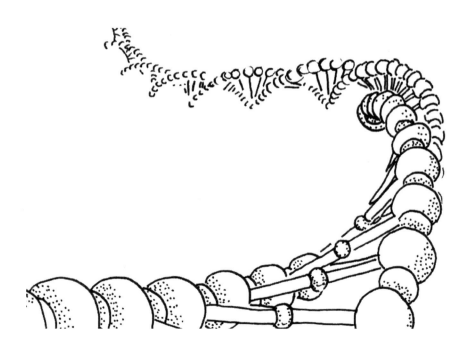

chapterTwo

Os Steward: Tennis Balls and String

> *I see the folks here at w2w as kind of the birth of a movement. A movement that's committed and anchored by passion, by knowledge, and even by accountability towards the end of paralysis.*
>
> Advocate Matt Rodreick

The conference begins with a brisk welcome from u2fp's executive director, Marilyn Smith. She's sharing words from a letter someone sent her that captures, for her, why we fight. *I've attended Working2Walk twice now, and I can't begin to tell you what a difference it made to me. I told my sister that I hoped before I die I would see my baby boy walk again. Unfortunately it didn't work out that way. Sepsis, fever of 108, irreversible brain damage.* We fight so that people will not have to endure what we did. That's the essence of it: we fight for the day when

nobody has to endure this. The tools we have are education, organization, and action.

She ends by introducing activist Bob Yant, who is the ultimate example of someone who's put those tools to work. Bob rolls to the microphone and thanks her; his job for the morning will be to act as master of ceremonies. The first thing on the agenda is a talk about a gene-based therapy from Os Steward, Director of the Reeve-Irvine Research Center. Here we go.

Os is a trim, healthy man in a dark jacket over a dark shirt, and he looks like he might be a documentary film director, or maybe a military person. He has that clear, focused air about him – a serious man who's used to giving orders and being taken seriously. He radiates authority, which isn't to say that he seems autocratic, because he also comes across as kind and patient. Above all and in spite of having a great smile, he doesn't seem like a salesman. This won't be a pitch designed to provoke unwarranted optimism.

The talk has a title that could use some unpacking: Regenerating Functions that Mediate Motor Function After Spinal Cord Injury. The first phrase – regenerating functions – means getting dormant or stuck biochemical processes to wake up and start doing their jobs. The second phrase – that mediate motor function – means the paper is about a particular set of dormant processes, namely the ones that either do or don't let muscles work. And the third phrase needs no explanation; this is about paralysis. So the title in plainer, more colorful English might be Goosing Clogged Gears That Turn Muscles On and Off in Paralyzed Rats. The talk will be an update on work that he and some colleagues published in late 2010. That work set off the equivalent of fireworks in the neuroscience world.

Os begins by recognizing Bob Yant's contribution to his own understanding of both science and the reality of life with a high cervical injury, in a way that is both gracious and heartfelt. He goes on (with some breaks for techy glitches related to whether he can use the pointer or get his slides to advance or have the lights lowered in the room) to credit his colleagues, including eighty-some people who work at the Reeve-Irvine Center. The photograph of them is a collection of young men and women, three of whom are in wheelchairs; it strikes me that this image might well contain the faces of the people who will cure paralysis. A colleague from a lab in Japan, Minoru Fujiki, gets special mention, both as the person who first got Os started on the particular area of research he's going to talk about, and as someone who's

> It's a big deal when scientists "publish." It means they've made important new discoveries in their labs. Being published is hard; it's evidence that researchers have convinced a group of their peers – who are also their competitors – that this paper describing their methods and results will move the whole field forward, and therefore needs to be in print where everybody can see it and, more importantly, use it to inform their own projects.

so dedicated to his work that he routinely flies to the Irvine lab from Japan, spends all day working, then flies home. Who needs hotels? Minoru not only does basic research but is also a practicing neurosurgeon, and he's here with us in the room today.

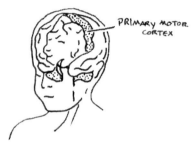

Before we get to Os's actual talk, we need a tiny bit of anatomy, for which we can thank Bob Yant. Bob likes to compare neurons and their axons to tennis balls with long, long strings hanging off them. The tennis balls are the neuron cell bodies, and the strings hanging off each one are axons. If you have a spinal cord injury, a lot of the tennis balls are in your brain, safe and sound, alive and well. The strings are what's broken, and what we need to restore. There's something else that we have to understand about those tennis balls and strings in order to make sense of what Os is about to tell us. The tennis balls – the neuron cell bodies – aren't randomly scattered in your brain.

They're sorted and functionally grouped in very particular ways, as are the strings – the axons – that descend from your brain and down into your spinal cord. The key to recovery of walking is the corticospinal tract, which is a particular collection of those strings. *Tract* is the name for a bundle of axons. Corticospinal tract axons start in the part of your brain called the *motor cortex*, which is how they got their name. No one has ever found a way to get these axons to grow past the injury site.

But Os and his team did that: they regenerated axons in the corticospinal tract. They got those axons to grow in mice a little – not all the way from above the injury to the end of the spinal cord, but just a couple of segments. He points out what everybody in the room already knows, that two segments makes a big difference in what living with an injury is like. (If my husband got two segments back, both his hands would work normally, which would change his life.) I'm a fan of this research.

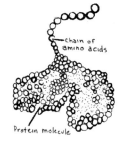

Here's the thinking behind their experiments. At some point in our lives, we stop growing. Right? We're small, and then we're bigger, and then we don't get any bigger. The reason is that our cells contain recipes that instruct them in what to do, and those instructions change depending on what else is happening in and around the cells. The recipes inside our cells are called genes, they're actually little chunks of the DNA strands that form your chromosomes.

Those little chunks of DNA arrange and manage everything that every cell in your body does. Everything. They do it by creating proteins, which in turn act as chemical cues to other genes, which create new proteins, forever and ever, amen.

This means that somewhere within your DNA, there must be a protein cue that calls out clearly to your nerve cells, "Hey! Don't grow any axons!" We know this because at some times during our lives (before we're born, for example) nerve cells produce axons in wild abundance, and other times (post-injury) they don't. The whole system is unfathomably complicated, which means finding a particular protein cue that nudges a particular gene recipe to execute a particular process and produce another protein which will be a cue to another gene, and so on forever, is – to put it mildly – quite a feat.

But that's what Os and his friends did. They figured out how to screen for the genes that are recipes for proteins that tell cells not to grow, and there was one special case that got their attention. I picture them like those guys who pan for gold, sitting in the sunlight with flat screens in their laps, shaking out the debris and watching for the gold to appear. The gold, in this case, is a gene called *PTEN* that produces a protein called PTEN. When they deleted *PTEN* (by snipping it out of the DNA chain, so to speak), the axons remembered

> Because genes are recipes for specific proteins, it's convenient to call them by the same names. In humans those names are always written in capital letters; the gene will be italicized and the protein won't. Hence, *PTEN* the human gene is a recipe for PTEN the protein it produces.

how to grow out of nerve cells again, like magic. Os says that a team from Children's Hospital at Harvard (now his collaborators, Zhigang He and Kevin Park) first figured this out in 2008. They were doing studies on the optic nerve, which is what connects your eye to your brain. If you cut the axons that make up the optic nerve, they don't regenerate, just as they don't in a cut spinal cord. It's an easier model to study than the spinal cord, but the principles are the same. The experiment involved first deleting the gene, then cutting the optic nerve, then watching that nerve regenerate.

Os then formed a team that included people from Zhigang's lab and people from his own, and they tried to make this work on rodents – mice, to be precise. The process was exactly the same: first delete the gene, then administer the injury, then test to see if nerves regenerate. He's careful to say that they gave their experimental mice the most gnarly kind of injury, which is not a transection (snipping it cleanly in two) but a crush (mashing it like an overripe strawberry). Complete crush injuries are considered the most complex and the least likely to permit recovery.

Their results were just the same: axons remembered how to grow, and they crossed to the other side. This is mice, though, right? It matters because one of the ways that mice differ from both rats and humans is that when they

get spinal cord injuries, they don't develop the kind of massive cavities in their cords that humans do. Labs use them a lot for gene-related work because they're simpler and the gene-related results transfer nicely to humans. Still, they're not always a good predictor for whether a procedure that's effective in them will ultimately be useful in us.

Os lays out the questions that are by now on everybody's mind, or at least those of us who've managed to follow the basic argument so far. First, did getting axons to grow actually make the mice with the crushed-strawberry cords regain use of their muscles? This might seem like a no-brainer, but it's really not. New axons have to find their way to the places where they hook up and form the appropriate communication chains, which there's no particular reason to assume is guaranteed to happen. They might just grow off into the weeds, figuratively speaking, and never connect to anything.

Second, how is this idea of deleting a gene before an injury ever happens going to help people, who after all aren't going to have their *PTEN* knocked out just in case they break their necks someday? Will it still work if it's done later – and especially much, much later? Bob Yant's injury is decades old, and he'd probably like to know if there will someday be a version of this treatment that can give him back the use of his fingers.

Finally, assuming there are good answers to the first two questions, how exactly are they planning to make this work for humans? Do they have something in mind already, and have they started testing to see if it might work? This moment right here is why I don't miss these conferences if I can possibly help it. Here's Os Steward, director of one of the most advanced and well-funded spinal cord injury institutes in the world, about to describe a way forward with real promise. And he's not a salesman.

Pause for breath.

He's got an experiment to tell us about that lands right where we want it to. He and his collaborator Gail Lewandowski figured out how to throw a wet blanket on the *PTEN* gene. It's called AAV-shPTEN, which is scientist shorthand for *adeno-associated-virus-short-hairpin-RNA-targeting-PTEN*. A virus is just a chunk of genes (RNA or DNA molecules) surrounded by a protein coat. Its only purpose in life is to push its genes into the cells of living things — which is what makes it a perfect engine for delivering new instructions to neurons. You take your virus, remove its natural genes/recipes, replace them

with the ones you want it to share, and send it into action.

This is what they did to create the wet blanket known as AAV-shPTEN, and it worked. When they injected that virus into rats after they'd been dealt their strawberry-crush blows, those rats got function back. The scientists had deleted the *PTEN* gene in the clump of neurons that controls voluntary movement. Without that gene telling those neurons that their axons were done growing, the axons grew. They formed communication networks, and the rats got function back.

* * *

Scientists have to get creative when they want to test for how much function their test animals regain, because they can't just tell an injured rat to lift its paw and hold it in the air. In the PTEN experiment they used a clever little box that rewards a rat for successfully using its forepaws to reach out and grab pellets. The box has six steps going down from the place where the rat is perched, and the idea is that it's pretty easy to reach out and scoop up a pellet from step one, harder from step two, and so on until it gets really challenging from step six. Os says that rats like this game, or at least they seem to like it, especially the eating part.

So, they found a gene with the *don't grow anymore* instructions. They figured out how to knock it down. They chose injury models, delivered the injuries, and cared for the animals. They came up with ways to measure functional return. Every single detail was planned and documented in as many ways as this very competent group of scientists could dream up. All that, and there's still one more new element to this experiment, which was the problem of how to deal with that cavity that rats and humans have but mice don't.

The solution they chose is called fibrin, which is a sort of glue – actually a protein that helps blood to clot. Lisa Flanagan, who works at the Sue and Bill Gross Stem Cell Research Center right next door to the Reeve-Irvine Research Center, had recently shown that salmon fibrin effectively and safely fills the cavity you get at a spinal cord injury site. So Os and his team used salmon fibrin as a friendly surface over which the regenerating axons could travel. Those axons did travel. They did form new and useful connections. Os has the preliminary data in graphic form, and we can all see that the injured rats who got *both* AAV-shPTEN *and* salmon fibrin were doing a lot better than the unlucky ones who didn't get either, or who got only one or the other.

So, is this it? Cure done? Hardly. It's very promising, and they're

going forward full steam – have in fact already done a number of experiments he didn't show us today that support the basic results. Os walks us through the obstacles and questions still remaining. They've chosen a particular time post-injury and a particular method of delivery in their rat studies, but they'll have to devise experiments that point toward what will work best in human beings. They also don't know how much recovery might be possible, which (this is my own comment, not his) will matter very much in the event that it ever becomes something to fight with an insurance company about.

The deletion of *PTEN* is only part of the answer; we know this because even in the relatively simple mice studies, those axons weren't growing more than a couple of segments past the injuries. There are other problems, including, probably, other genes that will need to be knocked out, not to mention the hostile environment at the injury site – what's called the glial scar. Finally, all the testing will have to be done on animal models that are closer in size to human beings. The difference between my cord and your cord is nothing like the difference between either of our cords and that of your average rat. A rat cord is like a strand of thin spaghetti. Ours are like long, thick ropes.

Os is not pretending any of this is easy or certain. He is, however, hopeful. And so am I.

chapterThree

Murray Blackmore: Gene Hunter

The hope is that I'll discover a gene or a combination of genes that, when applied to this chronically injured, dormant neuron, can kind of wake it up and enable its growth capacity. That's the hope. That's what I'm working toward.

Murray Blackmore

Okay. We hardly have time to shift in our seats. We just heard the famously cautious director of the Reeve-Irvine Research Center say that he's been successfully testing gene-based therapies that get motor axons to grow past the injury site. And, bang, we find ourselves being introduced to the next speaker. The man who gets up to follow Os Steward seems brave to me, even before he starts talking. He looks to be about thirty years younger than Os, for one thing, and for another, he's facing a crowd that just got some damned good news, some very hard-to-beat news. Unlike Os, he doesn't come across as a seasoned documentary film director; he's more like a new-ish professor – the kind who writes his own lecture notes, keeps liberal office hours, and makes a huge effort in class to be clear. No suit jacket, and his shirt is light gray and open-collared. Nerdy in an appealing way. We like him instinctively.

We like him even more after he starts his talk with a photograph of his own mother. Wearing a pink sweater, she looks like a standard grandma, except that she's sitting in a power chair with a sort of tray arrangement attached to the front. She's smiling, sharing a happy moment with the blonde toddler who's reaching for something on that tray. She has a cervical spinal cord injury herself, see, and what this means is that Murray Blackmore knows what it takes just to get through your day and stay healthy, much less travel by plane or car and check into a strange hotel to be at a conference like this. He's knocked out by us, he says – by the very idea of Working2Walk. He says that his mom was injured twenty-five years ago, and I sit there calculating that he must have been about ten or twelve when she got paralyzed, just the ages our daughters were when their dad broke his neck. Now I really like him, this Dr. Blackmore.

Because he and Os are working on similar strategies, some of what Murray says by way of explanation is going to be repetition of what Os just told us. This is okay. Many of the people in the room arrived not knowing

genes from apples, and the more times and ways we hear new information, the more likely we are to get our heads around it. These two didn't compare notes ahead of time; they just started with their own experiments and tried to make presentations in such a way that we could understand what they'd done and why.

Murray starts by talking about what paralysis is from the perspective of a nerve cell. The nerve cell can only send information if it has axons, and axons in the central nervous system (that's the brain plus the spinal cord) don't grow back once they've been damaged or cut. Why not? First, because the injury site in a damaged spinal cord is an environment that's hostile to axon growth. It's a scorched earth, salted ground scenario, though there has been a lot of effort made over the last couple of decades to figure out how to make it more hospitable. Second, because – as Os just reminded us – neurons in the brain and spinal cord don't regenerate their axons. Murray's focus for the last ten years has been on the second problem – the one that has to do with the neuron itself and not the scorched earth injury site.

He says that even if you could make the injury site friendly and welcoming to new axon growth, you'd still have to get axons to go ahead and extend and form new connections. The problem is that there's a built-in growth (more accurately, no-growth) program in our DNA, as we all just learned. Every neuron in the adult human brain and spinal cord has been given specific instructions that say, in effect, *You're done growing. Don't ever do it again.*

We're different from other species in this way; in fact it's more the rule in nature that axons do grow back. Flies, worms, zebrafish, tadpoles -- all can grow axons easily to replace broken ones, in the same way we that we grow new skin to repair a paper cut. What's strange is that we actually can grow new axons too, but only during development, before we're born, or in our own peripheral nervous system. Not in our brains. Not in our spinal cords. Why would that be? Murray's about to deliver a quick primer on DNA and genes, which turns out to be a great supplement to Os' talk, and which we need in order to understand what he's been up to in his lab.

You're made of cells. Every cell in your body (except red blood) has a nucleus, and inside each and every nucleus are your chromosomes. Your chromosomes are twenty-three double strands of a very unusual molecule called DNA. The same exact chromosomes

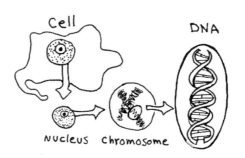

made of the same exact DNA are in every single one of the 60 – 90 trillion cells that make up your body. Your DNA is a sometimes called the blueprint that produced you, but it's really more like an elaborate cookbook that has the ability to reproduce itself, read itself, and carry out its own instructions. The obvious question is, if DNA is the set of directions telling each cell what to be, and if every one of your cells has the exact same instructions in its nucleus, why are you not an amoeba? How does the same collection of recipes sort itself out to produce hair and bone, neurons and hearts?

Good question. The answer has to do with how DNA works. It's a long, long chain made up of sections; the sections are called genes. Genes are basically do-it-yourself kits for building proteins. They're similar to recipes for making specific foods, but with built-in timers. *Make lemon cheesecake now, because the conditions are perfect. Don't make pea soup ever again.* The way the recipes get read involves a mediator, so to speak, called RNA; RNA is the machinery that turns a simple line of DNA code into a protein. And proteins in turn are catalysts, structural pieces, machinery, timers, us. There's a special set of proteins called transcription factors; their job is to bind themselves to the gene and either turn that gene on – get the RNA process started – or turn it off. Transcription factors are gatekeepers.

When scientists say that a gene is switched on, they're saying that its instructions to build a particular protein are activated; the turnstyle has been engaged. This is what Os meant earlier, when he talked about the *PTEN* gene making PTEN, its associated protein. What Os showed is that PTEN is one of the proteins that acts as a turn-off switch to axon growth. We know that, because when he switched off the *PTEN* gene – and thus stopped the production of PTEN the protein – axons suddenly remembered how to grow, and did grow.

Proteins are, in a sense, the keys to the kingdom. They construct and make up the machinery that runs cells, and they deliver the command and control signaling that determines when and how fast that machinery does its job. They're how we get new skin cells when we need them. They're how we get every kind of cell. And the switching on/switching off phenomenon explains why our bodies are made of such a variety of cells.

Murray explains all this in his earnest, careful, professorish way, clearly trying to give us enough information to follow along as he describes what he's working on at his lab at Marquette University. *This isn't just Biology 101*, he says. This matters because it means that, in theory, it should be possible to compare the genes in a switched-ON neuron (like you'd find in a developing embryo) to the genes in a switched-OFF one (like you'd see in a grown human being). If we knew which genes were the recipes for the grow-axons-now proteins, we might be able to activate them, effectively trading out the instructions that say, *Don't grow anymore* for new instructions that say, *Grow like crazy*.

He lays out his strategy. There are a few things that need to happen in order for a turned-off neuron to be re-booted and changed back into a turned-on neuron, and the first is to figure out which genes will need to be messed with. Os and his collaborators found the *PTEN*, but there's no reason to believe it's the only one. How to find the rest? You start by comparing those transcription factors in embryonic cells and mature cells, because they're the markers that show which genes are getting expressed. There are going to be three possible outcomes of that comparison. One is that some of the genes will be the same, which won't help us. What we want is the other two cases. We want genes that are switched on in the axon-growers and off in the dormant neurons, and vice versa. We're looking for what's different.

The problem for Murray becomes, unfortunately, that there are about a thousand candidates that meet that last criteria – a thousand possible recipes to test. Needle, meet haystack. This number temporarily knocks me over; inside each cell in my brain are a thousand protein recipes that had different instructions when I was a forming embryo. Fortunately, a couple of researchers from the Miami Project to Cure Paralysis, working on a completely different project, have devised a good search method to sort them out according to how they affected axon growth. Their idea was to "ask the neurons."

They just put the each gene into a neuron cell and watched to see what happened. This is not as simple as dropping an egg into cake batter, but for people in the business it's not that much harder. You start by buying the DNA with the gene in question expressed, something that's readily available to scientists. You then have to get that DNA into your test neuron and see what happens to that neuron. Murray does this on a large scale with rat neurons, ninety-six at a time. Next he uses an automated microscope and a set of mathematical algorithms to record the axon behavior.

When you weed through the mountain of data that results, what you see is that twenty of those thousand candidate genes tend to make the axons grow. Presto.

* * *

Okay, but all that gene manipulation and axon growth was happening in a dish, what scientists refer to as *in vitro* (it's Latin for "in a glass"). Can the same thing be done *in vivo* – that is, in a living animal? Sure. You do it with a

virus. Time for another quick tangent, because it's important to see how this works. Most of us use the word *virus* to describe the thing that makes us sick in flu season. A virus is actually a bit of a genius organism; it's got one simple job to do, and it's magnificently efficient at carrying out its plans. The task of a virus is to get inside a living cell and replicate itself as many times as possible, which it does by delivering its own RNA recipe to its host. The host recognizes the recipe and obligingly creates some special proteins called enzymes, which then assemble copies of the virus by the thousands. The copies go forth and do it all again, and again, and again, in every host cell that fits their requirements. In effect, the original virus functions as a super-efficient delivery system for RNA.

This means that once you know which RNA makes axons grow in a dish, you can put the axon-growing RNA into a virus, and then put that virus into the central nervous system of a living animal – and the virus will bring your RNA not just into one neuron but, eventually, into all of them. And that's what Murray did next, with each of the twenty different genes they'd identified in their lab dish tests. Of those twenty, they were able to show that one of them – the gene called *KLF7* – did indeed help axons to grow past the injury site, both when Murray's team delivered it by virus before an injury and when they did the injury first and the gene therapy later.

That's twice in this one morning that a gene therapy has been shown to do what used to be considered impossible – get corticospinal axons to grow past the site of a spinal cord injury. Murray shows us the data about regained motor function next. You'd expect that axon growth would mean better motor function, right? And his mice did indeed regain some ability to step across a ladder after their KLF7 treatment; it wasn't normal stepping, but it wasn't no recovery, either.

"I don't want to oversell this," is what he says next. They got a fraction of the axons to grow a few millimeters in a mouse, and saw some evidence of functional recovery. Good, but not cause for the purchase of dancing shoes. The question is how to build on that success. One obvious thing would be try combining the KLF7 treatment with the PTEN one . . . shouldn't the combination be more potent than each one delivered alone? Another would be to look again at the injury site and the problems it presents for regenerating axons; could they get more growth if they combined their genes with something that makes the injury site a little friendlier? Yes, again, and that's the direction they're going now.

It's all very encouraging, especially when you think about how quickly technologies and methods are being developed that make this even possible.

47

> The human genome is the name of the master cookbook. It has twenty-three chapters (chromosomes), each of which contains several thousand recipes (genes).

We didn't even understand the structure of the DNA molecule until 1953. The first creature to have all its genes mapped out was a tiny roundworm, and that was in 1998. A year later scientists had a map of all the genes on one human chromosome, and by 2000 there was a working draft of the entire human genome. Murray and Os (and thousands of others) are doing science that didn't even exist when they were in college.

What he's focused on now is trying to find more candidate genes. One way to do that is to sift through the DNA data sets that are common to all the regenerating neurons in all the species that are able to grow axons after those axons are damaged. Murray collects this data (as a hobby!) into a big excel file on his personal computer. He noticed that there's one gene that seemed to be common to all these axon-growers, called *SOX11*. He made a virus to express that protein, and – from preliminary data – it looks like this will turn out to be another axon-growing gene. For those keeping score today, that would be three of them. So far.

His last idea has to do with the relationship between cancer cells and growth. Cancer is all about unwanted growth, and gene therapy is all about creating molecular conditions for growth. Might there be an overlap? A safe, effective way to harness the bad that happens in cancer cells and put it to good use? Murray focused again on transcription factors – particularly a list of twelve of them, each of which has been shown to have some role in axon growth. He also collected a list of 210 transcription factors that have been shown to be involved in cancer. When he compared the list of twelve axon-growing transcription factors to the list of 210 cancer factors, eleven of those twelve were on the cancer list. *"That makes me really excited,"* he says. *"To check out these other 199."* He suggests that maybe it's a mistake to spend time on in vitro screening. Maybe we should just go right to in vivo testing and find those new molecular targets.

Does any of this have the potential to help repair chronic injury? We don't know. It's possible. It's definitely possible, and the tools are in place to find out.

chapter**Four**

SCI Regeneration Mothership: Is It Time?

> *We're talking about the researchers, the advocates, the family and friends, the funders . . . as stakeholders we really need to understand what motivates each other. We need to understand what frustrates each other, and we have to understand where the gaps are. Once we've identified those gaps and understand our resources, I think we can then maybe identify ways that we can overcome them.*
>
> Advocate Mark Bacon

Bob Yant is moderating a question-and-answer session that's meant to be a time for participants to ask for clarification from the scientists who just presented their slides. Sitting next to Os and Murray, he looks to be more than a full head taller than either of them, but that's mostly because his power chair seat is a lot higher than their ordinary ones. He's got a chest strap to hold him upright, and in his face is a quiet intelligence that's visible from across the room.

There aren't any questions from the crowd at first, so Bob asks Os to talk about the gene, *SOCS3* – something the two of them have clearly discussed before. Os gets up and goes back to the podium. He says to think about the pipeline, by which he means the idea that we all want there to be more things ready to test and hopefully combine into therapies that work. There needs to be a pipeline full of ready-to-go and promising genes. At Children's Hospital at Harvard, Os's collaborator Zhigang He has kept right on looking for more targets, using the same optic nerve scenario as in the early PTEN work. He found another one that has the same kind of results as they got with *PTEN,* and that's the *SOCS3* gene.

What's really interesting is that *SOCS3* is different from *PTEN* in terms of where it does its job.

Remember that RNA is like a mediator between the DNA recipes and the eventual cooking up of those critical proteins. It turns out that manipulating the *SOCS3* gene ramps up associated RNA production, and manipulating the *PTEN* gene boosts the translation of those RNAs into proteins. It's as if one of them (*SOCS3*) speed-reads the recipe and the other one (*PTEN*) speed-cooks the soup. And when Dr. He did the optic nerve experiment with both

those genes manipulated, he saw axon regeneration in the optic nerve boosted by a factor of ten. *Ten times as much regeneration as they got with PTEN alone.* The combination isn't additive, in other words. It's multiplicative.

And so of course, says Os, they're all gearing up to run the mice experiments and see if that result holds true in spinal cords as well. *You have to start at the beginning and get the mice to breed . . . we try romantic music, heh heh.* This little joke is so unexpected that it takes us an extra second to realize what he just said. Ahem. Anyway, he goes on to reassure us that they've got the animals with both genes manipulated now, and hope to start the first experiment early in 2013. We'll all be watching for news about that one.

Bob Yant inserts a pause here to make sure we all got what just happened. If, he says, PTEN gets two or three segments of axon growth in the cord, and if this combination with SOCS3 multiplies that by ten, that means . . . twenty or thirty segments. The human cord is thirty segments long. That would be the whole thing, the whole cord. Yikes. But this is very far from being done, right, and since there still aren't any questions from the floor, he asks another one of his own. It's about the *Encode Project*.

Encode stands for Encyclopedia of DNA Elements; it's a huge project in human genetics that's been underway since 2003, operating out of more than

400 labs and aimed at finding and characterizing every last functional gene in human DNA, including what used to be called "junk" DNA. What Bob wants to know is whether or not this project might point toward genetic switches that could turn on the axons.

So Murray stands up to say that he told a fib earlier – or at least over-simplified – what actually happens in the DNA → RNA → protein routine. One of the things the Encode Project has revealed is that that sequence only happens in the minority of cases. Most of the time, the RNA doesn't translate into protein, so the question becomes, what's it doing? It's not junk. It seems to be mediating complex layers of regulation; it's a vast and complicated switching system in the form of a sea of RNA molecules. This matters because Murray has technology and systems that give him insight into how to compare not just the proteins that are different between axon-growers and non-axon-growers (which is what he's been doing), but also how to compare the RNA sequences that are different. Maybe axon success or failure is happening at that level, too.

Next is a question from Dr. Jerry Silver, who's been in the audience listening to this presentation. He says congratulations to both Os and Murray for achieving axon growth in animals that got their genes manipulated after injury, something no one has ever managed before. What he wants to know is what kind of injury model Os was using in his PTEN studies. Also, can Os talk more about that fibrin/glue material he added to the injury cavities, which seemed to be critical to getting the axon growth and functional recovery.

Os says yup, only the rats that got PTEN deletion and fibrin got functional recovery; PTEN all by itself isn't going to get the job done. And the injury they delivered to the rats was a dorsal hemisection at the cervical level, not a complete crush injury like the mice he described in the first experiment got. Why cervical? It's the most common injury in humans. Why hemisection? Because at that level you can't do a complete transection; the animals couldn't survive it. It's going to be the model they use in their next round of rat studies, too, because it causes a big old injury-site cavity in rats, just as it does in people. And the cavity is going to need to be dealt with, so it's sensible to use animal models that work as much like ourselves as possible.

On the question of what fibrin is, it's a glue that's commonly used in surgery. Theirs was from salmon, for a lot of good reasons, including its ability to solidify. Os wants to leave it to Gail, who did the work with this, to discuss more when there's more time. (*We're late for a break*, he says. Yeah. And also, as my students used to say, *our brains are full*.)

There's a question from the audience about the pathway to the clinic.

When and how does a gene treatment get from a rodent model to a paralyzed person? Murray gets up and neatly splits the question into two parts. There's the science question, which is about the reason to believe that what works in mice would work in humans, and then there's the practical question, which is about funding and the FDA. Murray says that the science piece is clear: the genes he's been looking at are ancient and conserved. They're in flies and mice, which means that the gene's particular job is very likely to be just the same in humans. He defers on the question of what practical issues might be in the way of moving that therapy to people. It's not his expertise, but it's definitely something we have to figure out.

Next question comes from one of Working2Walk's most vocal and determined advocates, Martin Codyre, who managed to get here from Ireland last night in spite of the weather catastrophe on the east coast having caused so many cancelled flights. He has a pretty complete cervical injury himself, and he's asking, in his beautiful accent, who is in charge of making sure that people like Os and Murray talk to each other. Who's driving the ship? Couldn't they both be more efficient and work more quickly as part of coherent, managed effort? Who's in charge of the big picture?

"Nobody," says Os.

> Most of the recipes in our DNA are exactly the same as those of other creatures. In 2002, one analysis showed that 717 out of 741 genes on one chromosome were shared between mice and humans.

"That's what I thought! And that's a problem."

"It is." Os goes on to say that people in the field do of course talk to one another, especially at meetings like this one. But the idea of having a team of some kind guiding the whole project would require – here he gestures expressively – a huge amount of funding, which is just not there. "All of us are struggling just to float our individual labs."

Martin replies that sure, it would take very large amounts of money, but that money is getting spent anyway. It's inefficient to spend it in this isolated, competitive way. We need somebody to guide that ship and guide it globally. We need a ship that's not a Titanic – a ship that's actually going somewhere. Our community needs somebody who's doing that work, and if that were the case, the funding would be there. He's certain of this. We as a community need to focus on getting this solved. *We need an SCI Regeneration Mothership.*

Another voice from the audience is raised to say that he's seen a situation recently where a company got $18.5 million from a hedge fund simply on the strength of its science. There was good data based on sound science. Might it be time for all the scientists to get together somehow to demonstrate their capacity, and thus get all the funding that's needed to spearhead an overarching project to cure paralysis? Entrepreneurs have money, and they'd love to see something to invest in. Maybe we really can unite to fight paralysis. Maybe we should stop waiting for the NIH to dole out its grants and figure out how to get this done ourselves.

Bob says that we really have to take a break, but responds to the last suggestion by saying that exciting and encouraging as it is, this science is still at a basic research level. His impression is that investors are much more savvy and cautious than they were as recently as fifteen years ago. And then he closes by saying that he's been to maybe a hundred conferences since he was injured thirty-one years ago, and that it's been an almost surreal experience to sit here and have two scientists discuss their success at regenerating axons in the corticospinal tract.

And that's where we leave it, for the moment. Wheelchairs and walkers start moving for the doors at the back of the room, in a hurry to find the bathrooms and refill cups of coffee for the next session.

Injury Site Strategies

chapter**Five**

Damodar Thapa: Paralysis in Nepal

At its heart, the ADA is simple. In the words of one activist, this landmark law is about securing for people with disabilities the most fundamental of rights: "the right to live in the world." It ensures they can go places and do things that other Americans take for granted.

US Senator Tom Harkin, speaking on the 20th anniversary of the passage of the Americans with Disabilities Act

The speaker who leads us into the next few sessions doesn't work in a lab; Damodar Thapa is the head of physical therapy at the Spinal Injury Rehabilitation Centre in Sanga, Kavre, Nepal. Following him will be a pair of researchers who focus on the molecules that make the site of a spinal cord

injury so difficult to deal with, but before we get to them we have a chance to go a bit global. Damodar's presentation is a sort of back-to-earth moment, coming as it does while microscopic viral vectors bearing DNA code and dreams of a Spinal Cord Restoration master plan are still hanging just over our heads in this giant ballroom. It reminds me of how life with a spinal cord injury in the family can seem like an endless, random series of interruptions. *On your way to an important meeting? Sorry, this is going to be one of those days when your bladder spasms for no apparent reason. Go back home and change your clothes. Trying to get to your daughter's band performance? Not this time. All the parking spaces near the curb cuts are taken. You're stuck. She'll understand.*

So this talk brings the whole room back there, to the reality of how it is to live out of a wheelchair – except that the reality he describes might, in certain ways, be unfolding in another time and not just in another place. Nepal, as Damodar reminds us, is on the other side of the earth, landlocked between Tibet and India. Most westerners know of it mostly as a destination for extreme hiking; it's home to twenty of the highest mountain peaks in the world, including Everest, the highest of them all.

As you might guess, things are quite different for people living with paralysis there from the way they are here in the USA. Nepal's population is 27 million, which is just about the same as that of the state of Texas, but the landmass of Texas is almost five times bigger than that of Nepal. When people damage their spinal cords in the USA, the most common cause is motor vehicle crashes. In Nepal, half of spinal cord injuries happen when people fall from heights. Why so much falling? Damodar explains that most of these injuries happen in the countryside, where people climb trees daily to harvest valuable feathers, scale rock faces to collect honey, and mount ladders to work on building projects. His physical therapy center, just a few miles southeast of Kathmandu, is designed to meet the full spectrum of needs for injured people. It sounds very much like the trauma center where I visited my husband for the ten weeks he spent as an inpatient – with nurses, physical and occupational therapists, vocational training, social services, psychologists, and so on.

But whereas in the USA virtually everyone who sustains a spinal cord injury goes through that sort of rehab, in Nepal less than a third of them do. Of those who do go to rehab, quadriplegics spend between four and six months as inpatients, and paraplegics three months. (At about this point in Damodar's talk the giant screens at the front of the room light up with brilliant color close-ups of pressure sores. This never-ending threat to health always makes me angry, for some reason . . . probably just that it seems it should be preventable.)

In Damodar's PT center, patients do the same kinds of therapy that North Americans would be familiar with, but the outcomes, especially for the cervical patients, aren't nearly as positive in terms of independence and long-term health. The reason has to do with what's around them when they leave the hospital – a lush, beautiful country full of natural barriers to mobility, governed under a system that hasn't yet recognized the need for accessibility. His patients can't access the simplest care without coming to the capital city; his center is one of only two that are available outside Kathmandu.

The series of images illustrating what injured people face outside the doors of the hospital is daunting: homes with narrow doorways fronted by steps and surrounded by vegetation; a young girl in a wheelchair sitting marooned, looking seriously into the camera. There's a boy getting water from a well, as many people must do every day; how does someone in a chair work with that? Another picture shows a woman in a chair set on rocky, uneven ground at the edge of a field. She looks on while a family member collects a harvest from a garden. How does she contribute?

Damodar ends with a six-point plan that could be a beginning toward addressing some of the most pressing issues. When he came to Working2Walk last year, one of the things that stood out to him wasn't even on the program.

The sidewalks and streets had been modified so that people in chairs could use them; this seemed like something that could be done in Nepal, and within the last few months he's been helping to organize an effort in that direction.

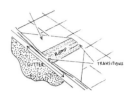

The rest of the plan has to do with helping people to see what's possible. The public needs to know what it means to have a spinal cord injury, and so

Photo credit: John Smith

there are teaching sessions in schools and other public arenas. There need to be more PT centers, and so he's established a satellite to his facility and hopes to create others. His staff needs to understand more clearly what the patients face when they leave the hospital, and families need information about what patients need – so he's set up a program that sends hospital staff out into communities to meet with families and form networks with neighbors. The last thing on his list is advocacy for rights of disabled people with the government.

What makes this talk so extraordinary is how it reminds all of us that we get to be in this room demanding faster science because of all the work that came before and all the people in chairs who did that work. Airplanes didn't voluntarily provide seating for disabled people. Hotels didn't voluntarily decide to make some rooms fully accessible. We get to be here at all because a lot of advocates refused to accept isolation and helplessness, and because professionals like Damodar backed them up. It's great to have him.

chapter**Six**

Ravi Bellamkonda: We Need Engineers, Too

My hope is to be able to collaborate with other researchers, generating the intervention strategies and especially the tools that my lab can provide to help move things forward

Ravi Bellamkonda

This talk is one example of the reason I decided I had to make this book. Sitting in the ballroom on the morning it was delivered, I was flatly unable to capture the sense of it in real time. I knew that for the people following the live blog, my posts about what Ravi Bellamkonda had to say probably seemed cursory and lame. They *were* cursory and lame, which means I wasn't able to deliver on what I'd promised to do – make the conference accessible to the people who couldn't travel. Yes, it was just a volunteer commitment, but Working2Walk is really a volunteer commitment for everybody involved, and I wanted to do it well. What's the point of going

to all the trouble of putting on a conference if its average audience member (me) can't really track what the speakers are telling us? I happened to walk past organizer Donna Sullivan later that day and she laughed, saying, "How'd the blogging go with Ravi Bellamkonda's talk, Kate?"

I just shook my head; totally not his fault, but it was frustrating. Too much missing context, too many unfamiliar words and phrases, too many undecipherable slides, and all of it presented *so fast*. To make this chapter, I got a copy of the video from that day and played his presentation at 60% speed, which made Ravi sound a little sleepy but also made it possible for me to comprehend what he was saying.

The overview goes like this: Ravi is both electrical engineer and biochemist, both maker of gadgets and analyst of central nervous system biology. His gadgetry expertise marries well to his neuroscience work, because the therapies that are going to combine and become a cure one of these days will need more than just knowing which cells or proteins or DNA blockers to deploy. Someone is also going to have to figure out how to get those things safely into the places in our bodies where they're needed, and that means the kinds of bio-engineered gizmos in Ravi Bellamkonda's extraordinary custom toolkit.

On paper and in person, he's impressive. A 1989 undergraduate degree in electrical engineering from Osmania University, one of India's most prestigious schools. A Ph.D from Brown University five years later, and a year-long postdoc at MIT after that. From MIT, luckily for us, he went to Case Western University and worked for a while under Jerry Silver, which is likely where he was infected with the cure-SCI virus. He says he thought at first that the problem was just a bunch of negatively charged scar tissue cells around the injury site. He would figure out how to neutralize the charge, and boom, regeneration would happen. (If only.) Today he has a lab at Georgia Tech, where a team of very capable scientists works under his supervision.

His talk is called, *Are CNS Axons Being GAG'ed? Alleviating CS-GAG Mediated Inhibition of Spinal Cord Regeneration.* *GAG* is a pun. In common usage, it means *stifle* or *silence* – but it's also the acronym for *GlycosAminoGlycan*, which is the name of a chain of sugar molecules that will be an important part of what Ravi has to tell us. The title is his clever way of suggesting that he's got a plan to prevent these molecules from doing what they're known to do: prevent axons from growing – i.e., gagging them into silence.

Some background here is important. Earlier this morning, Murray Blackmore talked about how there are two problems to be solved before there can be successful therapies to cure paralysis. Murray was focused on what he called the intrinsic one – the one that has to do with the built-in timers and instructions inside neuron cells themselves. The other issue is extrinsic – what I called the scorched earth, salted ground of the injury site. One thing that makes the injury site so unfriendly to axons has been known for more than a decade now – molecules known as CSPGs, which stands for *Chondroitin Sulfate Proteoglycans*. Sometimes you hear scientists talking in shorthand about proteoglycans. They mean these devils. *Proteo* is for the protein core, and *glycans* is another word for sugar. CSPGs are complex molecules with a protein core and some chemical sugars attached to that core like chains of jewelry. We're going to hear a lot about them, so it's important to recognize their various names: CSPGs, proteoglycans, protein-core molecules with sugar chains called GAGs attached – they're all the same thing.

We care about CSPGs because since the late 1990s it's been shown many times and in many labs that they're a key component in the molecular wall that axons have trouble passing through. They form a chemical/biological scar that repels axons. The good news is that there's an enzyme that likes to "eat" those proteoglycans; that enzyme is called chABC (for *chondroitinase ABC*). Again, many researchers in many labs have done animal model testing to show that chABC works. It destroys at least part of the proteoglycan scar, and that lets the axons grow, and that in turn has produced what scientists call "functional improvement" in injured lab animals. We call it walking.

So what's the problem? The problem is that the enzyme chABC degrades quickly when it's at human body temperature, but the human body keeps making those blasted scar-forming proteoglycans for many weeks after injury. And that means that a therapy involving chABC would have to be repeated over and over and over for months in order to work. The rest of Ravi's talk is about his team's efforts to understand very precisely how chABC does its cleanup job on the proteoglycans. Then he tells how he took that information and used it to work out the technical problems of efficient delivery and slow-release, both of which will matter very much for eventual human use. As a cherry on top of this sundae, he shows what happened when he used his clever delivery system to throw in some fertilizer on the (formerly) scorched earth injury site scenario. The fertilizer takes the form of what's known as a growth factor.

Here's what's interesting. None of this information is terribly new, at least not to the scientists or to the few super-well-informed advocates in the room. Ravi's team published the data about their excellent delivery system and its results three years ago. But it matters a lot that he's *here*, because the organizers want to make as many opportunities as possible for him to form relationships with the others in the room – both scientists and advocates. Somehow a collaboration of therapies, funding, engineering, political will, and personal courage must form. The idea is to bring the players together and get them talking until it does.

* * *

So. One important piece of what Ravi's team learned is that the chABC wasn't really digesting those proteoglycans whole; it was just eating the sugars and leaving the protein core of the molecules alone. And that meant that the sugars themselves were what was forming the scarred environment that kept axons from growing, because applying chABC was equivalent to removing just the sugars. And axons didn't have trouble getting past the remaining protein cores of the proteoglycans.

As Ravi put it, "*We have all sorts of tools to modulate and manipulate the environment after healing. We have nano particles, fibers, gels . . . but the key to success with any of those tools is we have to understand what biology we want to do to intervene. Absent knowing precisely what biology we want to do, the tools don't really help us. The journey in my lab is to understand well enough to make good use of those tools.*" Learning the precise biology involved comparing GAGs found in the intact central nervous system to those of an injured one. It turned out that there weren't just *more* proteoglycans in the damaged cord, but that there were different *kinds*. The structure of those sugar chains changed after injury, and knowing exactly how they changed is what made it possible to design an intervention that worked.

Sugars, as it happens, are sort of hard-wired once they're in place in our bodies; it's not like you can somehow slice them off the protein cores of those proteoglycans. This can be done in a dish, though, which is how we know that they're the culprits in preventing axon growth. A better strategy would be to prevent them from forming in the first place, and that means backing up to find the RNA recipe that creates the enzyme that creates the sugars. Remember that everything that goes on in our bodies is the result of some protein recipe getting activated, and an enzyme is just a particular kind of protein.

The most growth-inhibiting sugar has a marker that's not found at all in the uninjured spinal cord; it's known as a 4,6sGAG. The enzyme that's responsible for making 4,6sGAG is called C46ST. That enzyme became Ravi's target. They did an experiment to make sure they were on the right track by looking for the enzyme in both intact and injured spinal cords, and sure enough, C46ST was present in the latter but not in the former, just as you'd expect. This is what he meant earlier when he said that it would be important to have very precise knowledge about what was happening inside that biochemical scar.

Scar is the commonly used word, by the way, but it's probably not the best word to use. We all have a strong mental image of what a scar is: a patch of thickened skin that's different in texture and color from the skin around it. It seems plausible, with that image in mind, to somehow physically cut the scar surrounding the injury site away completely and allow new and healthy tissue to replace it. This isn't an option, though, partly because it would just be quickly replaced by more of the same. Ravi's tack was to figure out the specific properties in the scar that kept axons from growing past it and neutralize those. He could leave the rest of it alone.

Think about this . . . you know that proteoglycans have multiple sugar chains, you know exactly which type of sugar is blocking axon growth, and you know which enzyme builds that sugar – which means you can figure out which RNA has to be knocked down. All of that knowledge is years of work, building on years and years of other peoples' work in other labs. You have the tools to custom-make an siRNA (silencing RNA) that will fit the bill. How are you going to get that siRNA into the cells that need it?

Those cells are part of the smashed strawberry-contusion injury model; you can't just inject your siRNA in there without causing more problems. Ravi's solution is called a *liposome*. It's a nanocarrier – a 100-nanometer-long container that can be injected into your bloodstream; it's a sort of missile that will find its way to any place in your system were there's been a breach in your blood vessels and use that breach to get in. This works because the spinal cord injury site is such a place. There's been a breach, so the nanocarrier can get in. Inside this unimaginably tiny thing is the siRNA that will prevent those enzymes from being made, which means there will be no 4,6sGAG-type proteoglycans . . . which in turn means that the scar will be neutralized. The axons can grow.

* * *

And that was just the first half. The second half of this talk was about chABC, the enzyme he mentioned in the beginning that's known to attack proteoglycans by eating up all the sugar chains. Ravi had devised his elegant solution to the CSPG problem, but he hadn't completely abandoned the old chABC brute force approach. The problem with chABC, though, is that it's sort of like ice cream on a summer day; it doesn't stay fresh. Ravi's phrase to explain why was that it's *thermally unstable*. It loses 90% of its effectiveness within a few days of being injected, but those oh-so-unhelpful proteoglycans are being produced inside the injured cord for weeks after injury. For people, this would mean repeated injections. Even worse, in the case of the chronic cord any enzyme injected into the central nervous system gets washed out within hours.

So, in order for chABC to help humans, it would have to be delivered either in multiple injections or with a pump. It won't be practical until there's a way to keep it from losing its effectiveness at the temperature of the human body. Ravi says, "*We decided to fix that.*" You have to admire the confidence, right?

The solution they landed on involves a naturally occurring substance called *trehalose*. If you've ever been in the desert while rain was falling, you've witnessed the effect of trehalose. Plants that looked shriveled up and dead before the rainfall are suddenly plump and healthy afterwards. The trehalose in those plants' cells protects important proteins from the denaturing effects of heat, and as it happens it does just the same thing for chABC in the toasty temperature of the human body. Ravi's team gave the chABC enzymes a month-long bath in trehalose, then tested to see if those enzymes could chew up proteoglycans at 37 degrees centigrade. They could. They behaved just as if they were fresh.

The engineering problem became, then, how to deliver the new, improved, thermally stable chABC to the injury site. This time the answer was very tiny straws – like, says Ravi, the kind you drink Coke from, except they're very, very tiny, so it would take you a really long time (A joke! He told a joke!). The straws are only open at one end, so that whatever is in them drips out extremely slowly. That means that if you fill the straws with your thermally stable chABC, it not only stays fresh for a very long time, but the straws create a sort of timed release situation, so that the enzyme can eat up proteoglycans as fast as the body creates them.

When they tested this in animal models, they added some well-known growth factors to the straws, which is a bit like adding a slow-acting fertilizer to your vegetable garden. They invented a hydrogel to get the straws safely into the injury site, and what they saw after a single injection was that even after six weeks, the proteoglycans didn't come back.

After all that, did the axons grow past the scar? Did the animals regain some movement? Yes, and yes. Not all the axons, but some. Not full movement, but some. Ravi ends by saying that he wants to move ahead toward a cure by helping to generate new strategies and especially by providing the best delivery tools – liposomes and tiny straws and whatever else he can invent that avoids side-effects and enhances efficacy. I'm glad that so many people working on their own strategies are able to hear this talk, but I have to wonder how much effort is being duplicated. Sometimes this cure project feels as if we're trying to get an airplane built with some teams working on the fuel injection, some teams building the radar systems, and other teams designing the landing gear . . . but no central control to coordinate them. At least for this moment there are some team leaders in the same room.

chapter**Seven**

Jerry Silver: Patriarch

The reason I volunteer for Unite2FightParalysis is because of their philosophy of informing the SCI community about SCI research. There are two key needs. First we have to choose the best research and secondly we have to inform the community in language that they can understand.

Advocate Bob Yant

Even before he opens his mouth, you get a good feeling. Jerry Silver, Ph.D, professor of Neurosciences at Case Western University in Cleveland, Ohio, is like the favorite uncle who taught you to play chess when you were nine. He's warm and relaxed, with a full head of hair and black eyes that shine with energy and good will. He's going to be patient, and if you beat him in the end, you'll know it wasn't because he let you . . . he really wants you to learn. Jerry starts by saying that Unite2FightParalysis is one of the most

important foundations in the country for advocacy and education. He loves them, and he's glad to be here – so glad that he's doing two talks this year, one today and another tomorrow. He says with a half-wink, *"And I'll do more if you like."*

Today the formal title is "**Functional Recovery after SCI with the Use of Injectable Peptides**." Already this is good news; *functional recovery* means that something they did worked; *the animals got better*. *Injectable* means that the drug was delivered in a normal, simple way: with a needle and syringe. His first slide contains an image from Ramon y Cajal's textbook, and its style will be familiar to everyone who's ever studied the general topic of central nervous system regeneration. As a boy in his Spanish village more than a hundred years ago, Cajal wanted to become an artist. At his father's insistence, he went in a different direction and eventually won a Nobel Prize for his work in exploring the structure of the spinal cord and the cells that make it up. His original drawings of neurons and axons in the injured cord are eerily perfect representations of what he could barely see; we only know how exact his instincts for the structures were because we now have microscopes that cost as much as houses.

Jerry uses the drawing to illustrate the problem. When axons are cut, they seem to shrink back a bit and then curl up into useless little stubs. Cajal thought that they just eventually degenerated and got removed by our clean-up cells as if they were so much debris, in the same way that any debris in the body gets handled. This, Jerry says, is not true. Cajal was wrong. They're all there, all the broken axons, just sitting in the white matter of the cord, forever. The question is, why?

I've said earlier that the environment of the cord is hostile. Scorched earth. Salted ground. Nothing can grow. Ravi Bellamkonda just delivered a seminar about proteoglycans and the particular sugar chains on them that prevent axons from growing. Jerry's question is not for the moment about the proteoglycans, but about the axons themselves and this odd, unexpected behavior. If they're not dead, why are those axons just sitting there? Why does a living axon stall in place in the white matter – a behavior that's deeply unnatural, from the point of view of the axon? Their endings should be in gray matter.

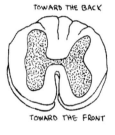

Argh. We need a quick trip around the spinal cord to know what he's talking about. If you slice through a spinal cord like a sushi roll at a 90-degree angle and look down at the stub you've exposed, what you would see is something like the image. The cross section is round. In the center is a butterfly-like gray shape, surrounded by a different material that appears to be white. To get the hang of this structure, I think about an individual neuron up in the brain. It's sitting there with its single axon extended down into the cord. That axon is coated in a white substance called *myelin* that acts like insulation on a copper wire. It's just one of millions that form long bundles; every single axon in those bundles is looking to get its terminals into that gray matter. It's their destination.

Inside the gray matter, they're home. Waiting for them are what they've been looking for: the *dendrites* from cell bodies of other neurons. Hello! *Dendrite* comes from the Greek word for tree; its name captures the branch-like look of that part of the neuron. The axon terminals approach some likely dendrites and

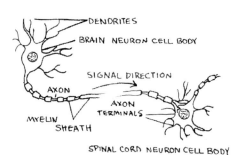

start making connections. This is how we make our muscles move. Axons from nerve cells in the brain find dendrites from nerve cells in the cord. After that the nerve cells in the cord send their own axons out to the muscles. The nerve cells inside the gray matter of the cord are sort of like relay runners. This is why it makes no sense for a live axon coming from a perfectly healthy neuron in the brain to just hang out endlessly in the cord's white matter, where there are no dendrites. Why would it do that?

Jerry switches to a slide that's bright with preschooler colors. There are four rectangular frames on this slide, and each one is a simple line drawing: blue skies, yellow suns, green hills. In each frame is also the image of a neuron cell, with its fat body, gray nucleus, and long tail of axon. This slide is about the possible reasons that axons don't grow back once they're damaged in the spinal cord. One or more of these little frames is an illustration why paralyzed people stay paralyzed. They're the details behind "scorched earth."

The first possibility is at the top left; in this image the neuron cell body is sitting on a green hill, and its axon reaches across the land toward an impenetrable red brick wall, where its stub lies hopelessly, apparently having smashed and fallen. It would go on if it only could, but the wall prevents it. This is the glial scar hypothesis. That wall is made of proteoglycans. At the top right, the frame shows an axon draped over a beach chair under a bright yellow sun, taking its leisure and enjoying a foamy drink. The beach chair represents something that the axon likes; in this representation, growth doesn't stop because something *prevents* it. Growth stops because the axon has found a comfy spot, a sort of oasis, and it prefers to stay.

The bottom left frame shows our axon stopped at the rim of a deep gorge, through which shallow blue water runs. The axon could maybe grow across to the other side, but the gorge is deep, the gap wide, and the cell body too old and tired to make the effort. The gorge represents the cavity in the lesion. The river, too far away at the bottom, is the absence of needed growth factors – vitamins that the tired neuron cell is lacking. The last frame shows our axon at the edge of wide blue lake. Swimming in that lake are sharks, crocodiles, and evil-looking red monsters; these all represent inhibitory molecules inside the glial scar.

Jerry says that he first drew these cartoons twenty-five years ago, and that he had no idea it would be the one with the beach chair that held the key to what's actually going on with those trapped axons. Great. I'll henceforth have to revise my personal metaphor now for what's going on at the injury site. Not only is there a scorched earth scenario preventing the axons from growing past and on to where they could make new connections, there's also a secret pleasure room, so to speak. The axons aren't just stopped, they're lured into staying.

* * *

We're going to go back to talking about proteoglycans, but not just in the context of that complicated injury site environment. There are CSPGs in all kinds of places where the body wants to prevent growth, including the uterine *stroma*, which is a kind of supportive network of tissue that surrounds every human uterus. What are proteoglycans doing there? The same thing, it turns out, that they're doing at the injury site: preventing an invasion. A uterus is made to contain a placenta, which is the living container for a growing fetus. If the placenta were able to grow through the wall of the uterus and invade the mother-to-be's body, it would kill the mother. Every time. *"We owe our existence to those proteoglycans,"* Jerry once said in an interview with the Reeve Foundation's Knowledge Manager, Sam Maddox. *"They're potently inhibitory. They're like a guard rail."*

So what exactly are they doing in the healthy spinal cord? They're forming a protective little barrier around each synapse, each place where the axon terminals and dendrites come together in their endless task of running messages between brain and body and back again. Once those connections have been formed, the last thing we want is to have them fail, and the proteoglycans are very effective at making sure all our hard-won learning stays in place. As Jerry says, *"We don't want the brain to change every day, right?"*

Right. Jerry's lab did an experiment years ago where they put mature sensory (not motor) neurons right into a spinal cord with an injury in it. They saw those neurons grow happily and abundantly up and down the cord, except in the injury site itself. The axons will grow right into that site and then just stay there, and they'll do it even in chronic injury models – which means that even a long time after injury, most of the cord is happy to accommodate new growth. But not in the scar; that territory is like fly paper for growing axons.

On the screen now there's a slide showing a time-lapse movie of sensory neurons growing in a dish coated with a substrate made of a growth-promoting matrix molecule called *laminin*. What we see is a pale gray background with some blobs on it, dark and irregular; they seem to be attached to one another by something like barbed wire – long threads with tiny spikes coming off. At the ends of the threads are small amoeboid-like creatures that appear to be reaching for something, pushing outward with arrays of little feelers. And then we see that some of the tiny spikes are also doing this pushing out thing . . . these are axons, growing, extending themselves. But if you could paint a sharp-edged stripe of proteoglycans into that dish, what you would see is all those little axons turning away. We do see it, right on the screen, because that is exactly what they did. And when they added some chondroitinase, the axons went forward again.

Remember the child's illustration slide with the proteoglycans appearing as a tall, solid crayon-red wall? It turns out that *in vivo* after injury they're not arranged like this at all, but instead are more gradually shaded, like a watercolor of a wall. At the rim of the injury site, they're thinly scattered, but at the center they're more of a solid mass. In Jerry's lab they figured out a way to reproduce this gradient situation in a dish, and that's partly how they finally understood that there really was a beach chair. In a gradient of proteoglycans, the axons enter from the low side and get lured into the depths of the gradient. There really is not *just* a steep wall, and a deep dry gulch, and a lake full of hostile creatures – in the body of a living animal, in places where proteoglycans are in the highest concentration, there is also a strange little oasis, right in the middle of the injury site.

The beach chair – the thing in the oasis that makes those axons want to sit down and relax and enjoy the rest of their lives – has to do with a tricky interaction between several kinds of stem cells and the molecular soup they make. The stem cells are called *oligodendrocyte progenitor cells* (OPCs). Oligodendrocyte comes from Greek words that mean "cells with a few branches." They're the cells that make the myelin insulation for every one of your axons. *Progenitors* is the word for cells that are waiting to turn into something specific. I think of them like vice-presidents: they can only become presidents, and they're around in case something happens to the existing presidents. Progenitor cells

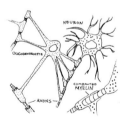

are like that, but they have a feature that vice-presidents don't: they can endlessly reproduce themselves. That's what makes them stem cells.

Anyway, the molecular soup in the injury site is a mixture of both growth-promoting extracellular matrix molecules, including the one called laminin, and lots of proteoglycans. It turns out that this particular mixture of molecules – especially the proteoglycans – entices the re-growing nerve fibers to get incredibly stuck upon the surfaces of the stem cells, so tightly stuck that they can't move. What's actually happening is that the growing axons are forming what feels to them like synaptic connections. Axons are born to make synaptic connections. It's their job. These are not real synapses, so no information is being sent from brain to body, but from the point of view of the axons, when they find these cells in the lesion it's apparently close enough. They think they're home.

What's bizarre, says Jerry, is that the progenitor cells just love the lesion environment. They proliferate there after all types of central nervous system injuries. So we have a bit of a marriage made in heaven at the injury site. Lots of OPCs, lots of proteoglycans, and axons fooled into thinking they're doing their job. Interestingly, it has recently been discovered that OPCs actually do make a kind of synapse with neurons *normally* throughout the central nervous system.

Synapse comes from a Greek word that means "to clasp," but it's not a good description of what actually happens. There's really an impossibly tiny gap between one axon and another, about as wide as a human hair split 20,000 times. That gap is the synapse. No clasping is involved.

Their job in the undamaged brain is to sense activity in the axons, so that if a resident myelin-making oligodendrocyte might die, the progenitor cell nearby can sense this and rapidly differentiate to take the place of the dead cell. The vice-president transforms itself into a president. "Sensing activity" in nearby axons is the OPC version of a synapse. Pretty clever system.

* * *

So, we know what's making those axons linger in the injury site instead of growing out beyond it. They think they've formed synapses. I said that a synapse is a super-microscopic gap between the axon tip and the nearby dendrite; what I didn't say is that the super-microscopic gap isn't empty.

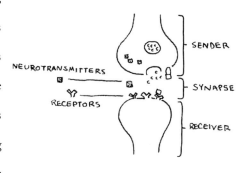

During an actual synaptic event, one side is sending out molecules called neurotransmitters across that gap. On the other side of the gap are waiting protein molecules called receptors; the axons trying to grow through the injury site have special receptors at their growing tips, and those receptors are highly adhesive. Interestingly, those super-sticky receptors increase as they see more

proteoglycans and "sniff out" the possibility of synaptic connections. When the receptors find them in abundance, they bind themselves tightly to any nearby proteoglycans. The proteoglycans in turn bind the axon receptors in such a way that they get switched to a sort of ON position. Thus, the ends of the axons, instead of growing their way through the injury site and finding real dendrites to hear their messages, just sit there, alive but stuck, like flies on flypaper. Instead of forming a link in the communication chain, they just activate the local OPCs in the middle of the injured white matter and sit there uselessly in one place.

Okay. The thinking for the experiment Jerry is about to describe goes like this: maybe you don't have to get rid of the proteoglycans at all. If what's keeping the axons stuck is the receptors – the axons' own reaction to what they see as potential synapses – then you might free the axons by working on that end of the equation. What Jerry's team figured out is the exact nature of the special adhesive receptors on the axon tips in the injury site. If you could somehow neutralize those guys, you might destroy the axon's ability to see the flypaper zone. You might prevent this crazy stuck-in-white-matter behavior. And that's exactly what two of Jerry's young partners (he shows their picture, calling them geniuses) managed to do. They turned those receptors OFF so the axons could de-adhere. The turn-off switch is a custom-made *peptide*, which is a simple chain of amino acids that's found in mice, rats and humans. In Jerry's lab they built two peptides, one for each of the two receptors; they call the peptides ISP and ILP. They put ISP and ILP on a kind of biological shuttle bus that carries them safely and easily into cells.

Once they had evidence from their so-called "scar in a dish" gradient culture experiments that these peptides actually worked, they went on to live animals. The *in vivo* experiment started with adult white rats and a severe contusion injury – not a clean hemisection, but a more complicated, smushed-cord injury site, like that which occurs in humans. They sorted the rats into four groups: a control group, (which got plain old saline), an ISP group, an ILP group, and an ISP + ILP combination group. Starting one day after injury and then every day for seven weeks, each rat in each group got a simple under-the-skin injection in their backs just above the lesion. It was a double-blind study, meaning that nobody working with the animals knew which treatment which animals got.

What happened is evidence – as if we needed any more! – that the injured spinal cord is a scene of great mystery and complexity. The controls, of course, did not get better. Neither did the combination group or the ILP group. In the dish experiments, the ILP and ISP both worked; axons became less stuck and grew. In the combination treatment the axons became too loose and fell off. Axons apparently need just the right amount of adhesion to grow. For reasons yet to be determined, only the ISP worked in the animals. In the lucky ISP group of injured animals, there was a lot of recovery – but not always the same kind in each animal.

The scientists were measuring three things: over-ground walking, grid-walking, and urination. About half the ISP rats were "responders," meaning they got significant return of at least one of those three. There were responders who got all three, just two, or only one. For instance, some got recovery of urination and no improvement in walking or grid-walking. Importantly, thirteen of fifteen animals recovered at least one of the functions. Some responders got almost full recovery of all three behaviors! At this point it's not clear what separates the responders from the non-responders. They don't even know for sure *where* the peptide did its work; it could have been anywhere in the rats' central nervous system, because the injections weren't targeted for a particular place. The lab will have to do anatomical analysis to understand what happened.

(Note: Jerry reports that very new work following the symposium is showing that the peptide is stimulating an unprecedented degree of sprouting in a very important descending system of axons that make a neurotransmitter called serotonin. This particular brainstem

projection to the spinal cord is very important in controlling locomotion as well as urination. Very interestingly, the amount and pattern of sprouting varies greatly between animals, and this could help explain the variability in recovery.)

What they also know is that this peptide, which is simple to make and simple to deliver, worked better than anything they've ever tried, including chABC. Jerry's lab is already devising experiments that will use their peptide in chronically injured animals. These will be combination-therapy trials that put together several of the treatments we know have shown results: peptides, chABC, and intraspinal micro-stimulation. The peptide treatment that was the heart of this talk is just one part of what will be, someday, a whole array of treatments that combine to make spinal-cord-injured people well.

chapter**Eight**

Injury Site Obstacles: Are We There Yet?

If the community wants to see collaborative projects, I think the money that flows to the scientists should have strings attached, and I think that should be a string. I think that collaborative projects should be funded, and the individual lab projects shouldn't be. So that would be a very direct way to force collaboration.

Murray Blackmore

Bob Yant and the last three presenters are sitting together at the presenters' table, ready to take questions on what's just been said. A man from the audience wants to know Jerry's thoughts about combining his peptide with rehab. Jerry says that it's a really important question, and that in his lab they're doing exactly that experiment right now, even as we speak.

What's he talking about? Most people in the USA (though not in Nepal, as we just learned) do some kind of rehab after a spinal cord injury. For my husband, who has a c6 incomplete injury, rehab turned out to be a three-part

process. First was about seven weeks of daily inpatient work, every bit of which was aimed at making him independent in his wheelchair once he left the hospital for good. The goal of the hospital team was to let him go home safely, not to recapture the use of any muscles that were still firing.

Second was about eighteen months of expensive and frustrating outpatient rehab at a standard hospital physical therapy unit, which was aimed at first building up core muscles, then being hauled to his feet to simply let his body remember how not to faint when upright, and finally – wearing one full leg brace, a gait belt, and with therapists on either side and in back of him – to take excruciatingly tiny, slow steps. He had great trainers there, without whom it's extremely unlikely that he'd have managed to walk at all.

He was among the lucky in that he had enough surviving axons after his injury to make that possible. But what Jerry is talking about that's happening in his lab has to do with the third part of my husband's rehab, and that was his participation in a clinical trial that tested suspended gait training. Here's what happened: he went to live in Gainesville, Florida for ten weeks. Every day he had a two-hour session with three graduate students and one post-doctoral student in their lab at the University of Florida. They would fasten a harness that hung from an overhead beam around his middle while he stood on a treadmill, then adjust the tension in the straps so that he didn't have to support all his own weight.

The post-doc student would operate a computer that was recording information from devices in the treadmill: stride length, impact, weight shifting, speed. Grad Student One would stand behind my husband, straddling the moving belt and holding his hips to keep his body square and stabilized. Grad Student Two would sit on a low bench, facing my husband and physically placing his floppy right foot on the moving belt in a natural rhythm, over and over and over as my husband "walked" along. Grad Student Three would be standing by, watching and trying to recover from doing the Grad Student Two job, which was physically very challenging. I know this, because they let me try it.

My husband's job was to imagine himself walking normally at close to his old speed and try to do it. If you think about how it looks when someone walks behind a walker, you'll see how far that activity is from ordinary walking. Behind a walker, the person can't get a natural rhythm, and it turns out that this is the most important part of retraining walking in someone who has just enough axons firing to make it possible. *Rhythm*. Walking is a rhythmic activity that involves getting many muscle groups to all sync smoothly together. It involves arms moving in opposition to the action of the legs, which was the whole point of suspending him over that treadmill.

He came home from that place after ten weeks, able to get around quite nicely with an ordinary cane.

* * *

Okay, back to Jerry. His answer to the question about combining his peptide with rehab was that it would be very important, and that they have a treadmill for their treated rats. I picture grad students placing those tiny feet in the correct rhythm, which seems impossible . . . but then so does figuring out the exact receptor that fools an axon into thinking it's forming a synapse and neutralizing that receptor so that the axon goes ahead and grows. Jerry says that what they're now trying to do is see if they can use rehab to get *all* their rats to respond, and to get the ones that are already responding to get even more function back.

Ravi asks Jerry if they've tracked what exactly those peptides are doing once they get into the rat's systems. Do they know what percentage of them actually make it into the rat spinal cords? Jerry says no, they don't know. They think the peptides are going everywhere, and they're doing tissue analyses to find out.

Bob wants to know if Damodar is sharing information from this conference with his patients, and Damodar says he is. His patients are very curious about what's going on in the west, especially with respect to a cure. It occurs to me that the reason Damodar belongs in this set of speakers might be that Nepal – or anyplace where the outcomes after injury are so grim – is a sort of metaphor for the scar tissue that Ravi and Jerry are addressing inside the cords of their injured animals. Damodar is struggling to get information to proliferate into a broken system, where the barriers aren't proteoglycans but rather poverty, illiteracy, and lack of information. Later, when a questioner asks him why only a third of people in Nepal get any kind of rehab at all, that's essentially what he says.

There are questions about why Jerry is working with sensory neurons instead of motor neurons. What's wrong with adult motor neurons? He says that they did their in vitro work with adult sensory neurons for a couple of reasons, the main one being that motor neurons are very, very difficult to get out of the adult nervous system. Why not use embryonic motor neurons, then? Because they follow different rules than adult cells, and one of those has to do with the receptors that the peptides get rid of. Embryonic motor neurons don't have those receptors, so you wouldn't learn anything useful. In any case, the peptides obviously did have an effect on motor neurons in the rats that were "responders."

We talk about Ravi's 30-day thermostabilized chABC; is that going to be a long enough window? Is he going to have to figure out how to keep it fresh and functional even longer? This question is really about how long those blasted proteoglycans keep getting produced in the damaged cord. It depends, they think, on the kind of injury.

Ravi wants to know whether the blood/brain barrier stays open in the chronically injured cord. This is important because his nanocarriers, remember, are engineered to find tiny breaks in the vasculature so they can slip into the cord with a minimum of fuss. Jerry thinks that the barrier closes within a few weeks for large molecules, but stays open for a long time for smaller ones. Ravi wants to do simple IV injections of these carriers, loaded up with whatever good stuff makes sense. He's doing experiments now that will look to see if the gap is as wide as 100 nanometers, and expects to have that information in about a year. The nanocarriers could be loaded up with not just chABC, but also with growth factors and/or siRNA. Jerry wants to do a combination that starts with chABC and then adds his peptide in a separate injection once the axons start to grow.

Around this time Os, who has been sitting in the audience, asks for the microphone. He has a comment, apparently provoked by this talk of combining therapies and a couple of polite-but-insistent questions from the audience asking *when, when, when* is even one piece of this going to be available for human beings? This is what he says:

I don't have a question, but I wanted to make a comment, going back to one of the earlier comments about why we scientists don't all get together in a kind of Manhattan Project kind of thing. Let me lay down a challenge here. Think about what that would mean. You've heard us talk about a variety of strategies here, and we're all saying we need to do more experiments and find good timing and dosing and move to larger animals to scale this up . . . let me just throw a number out there – a challenge. I think one billion dollars would be a conservative estimate about what it would take to make it possible for all of us to stop writing grants every three weeks and take our research to the practical level.

Martin raises his hand. *My answer to that is, okay, but where's the plan? Where's the cohesive globally cooperative plan for that? People aren't going to just write a check unless there's an Oppenheimer. Who is that? If that's there, I don't think money is the issue.*

And Os closes this long and remarkable morning's conversations with this.

There's not going to be a single person who stands up to say, "I'm gonna lead this!" Because we all have different attitudes and ideas and approaches that are all reasonable. But if there was a possibility of coming together and funding a really imperative group effort, we could come together and reach a consensus opinion of what to do. But you have to have the money to create that vision . . . and then we could do it.

I want to add my own coda here. A billion dollars sounds impossible, until you remember that we're talking about the global community. What's a billion dollars? It's a small fraction of the annual cost of care and lost income for the hundreds of millions of people living with paralysis.

Among Friends

chapter**Nine**

Advocacy Is the Reason We're Here

Somewhere, something incredible is waiting to be known.
Carl Sagan

Included in the price of admission to Working2Walk is lunch, served at the tables where we've been hanging out for the last few hours. A uniformed wait staff hustles into the room hoisting giant trays full of food — today it's chicken and mashed potatoes, salad and carrot cake, which I get to enjoy while chatting with Os Steward about his experience of the morning program. (He seems very energized, but I have a feeling that might just be his natural state of being.) Once the plates have been cleared away again, Marilyn goes back to the podium to set up the next few speakers.

She starts by talking about our friend Geoff Kent, who was paralyzed in a skiing accident at Steamboat Springs in 2007. Geoff came to Working2Walk

the very next year and left determined to figure out how to help push the cure along. That determination turned into a nonprofit organization called *SCI SUCKS*; if you were to google those words, you'd find his website. My favorite thing about it is the subtitle. (*SCI SUCKS: People tell us to change the name, but then we wouldn't be telling the truth*.)

Geoff's first goal was to raise awareness about spinal cord injury, and maybe even make some money for research. He'd already been interested in wheelchair racing, and so he talked six of his able-bodied friends into doing the Chicago Marathon with him as a fundraiser. They ran/wheeled their first race in October 2008, and astonished themselves by raising $50,000. In 2013 there will be SCI SUCKS teams raising both money and awareness by doing marathons in Chicago, Naperville, and Los Angeles.

The success of Geoff's fundraising efforts led directly to a difficult question: *where should the money go*? How do you even start to understand what would be a good use of research dollars? Marilyn says that it was Geoff's idea to take the step that she's announcing today. Together with SCI SUCKS, Unite2FightParalysis has formed two new entities: a Scientific Advisory Board and a Research Advocate Committee. As the 2012 Working2Walk guide put it:

"*The two groups will work in tandem to review, and recommend for funding, research projects targeted toward repair of the chronic spinal cord injury. The scientists and global community advocates that constitute the Boards have been carefully selected to produce transparent, critical analysis of the most promising cure science for the SCI community. This will better enable individuals and organizations to direct charitable dollars to the best labs."*

It would be hard to overstate the importance of this move, and though I stay in my seat cranking away at the live blog, I really want to stop the show for a minute and make sure people heard what she just said. A highly qualified scientific advisory board focused on **nothing but curing the chronic injury** as fast as humanly possible, collaborating with an advocacy group that will keep that mission front and center! Such a thing has never existed, anywhere, ever.

Effective advocacy has to be well-informed, just as effective science has to be well-funded. It's a sad fact that there are always people looking for money to work on "cures" and even treatments that have no solid grounding in science. That's not just wasted money; it's lost time. Money spent on sketchy projects and therapies actually slows down the pace of getting to a real cure for chronic injury. These two boards will help to put a stop to that.

This part of the program is meant to display a few examples of effective advocacy. Three very different people, all of whom live with paralysis, are on

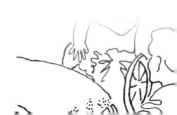

the agenda as examples of what can be done. (I recommend, as always, that you should spend some time watching the videos of these presentations.) What follows here are quick descriptions of the speakers, along with some of the highlights from what they said.

* * *

Bob Yant

I went to one of the early Society for Neuroscience meetings. This could have been in about 1982 when it was really sort of forming. At one point in the meeting they said, "Anybody that wants to talk about spinal cord injury research, you can just go over in the corner over there." And this is really true: there were about fifteen people that went. And that was it, at the time. In the whole world.

About nine years ago I maybe got a little frustrated that I thought maybe the research wasn't progressing at the speed I thought it should. So I sort of stepped back and, like – you know how you take a piece of paper and you draw a line through the middle, and what are the pros and cons? In my case it was like, what have we accomplished, and where do we still need to go?

I thought I would go a step further than that and ask scientists what we really needed. So I polled twelve of the top scientists that I could find. I asked them. I just said, "We're paralyzed. What do we need to get out of our wheelchairs? What do we need to

---

Bob Yant broke his neck in a diving accident in June 1981. He was a board member of the Christopher and Dana Reeve Foundation before it was even called that, serving in that role from 1982 to 2011. He's just accepted the job of chairing Unite2FightParalysis' new Advocate Research Committee. And one more thing? He's personally raised more than $10 million for the cure. So far.

accomplish this?" And I got a unanimous answer that came back to me. And the answer was, "You need to regenerate this tract of nerves, called the corticospinal tract." What we saw this morning was — and I told you it was surreal for me; I really almost felt like it was an out of body experience to see Dr. Steward and Dr. Blackmore showing they actually are regenerating those things for the first time.

* * *

Roman Reed

My whole life had been physical movement. I'd been captain of every single team I'd ever been on — football, baseball, basketball, swimming, soccer — you name it. My whole life was movement.

But on September 10, 1994, my whole existence changed. I went from being able to do everything to . . . nothing. In a single second. Doctor came into my room three days after the fact and said, "Roman, you'll never move your arms. You'll never move your legs. You'll never father a child."

I said, "Eff you. Get out of my room." I didn't accept that then and I don't accept that now.

My dad and I went to Stanford, and we went to the medical building. And we looked up spinal cord injury, and there's all these books, and we spent about $300, and we bought all these books. And we were determined. We were going to find a cure. And we read

Roman Reed was a star athlete when he broke his neck playing college football in 1994. He and his dad – the phenomenally tireless advocate, Don Reed – became the core of a group of Californians who moved the mountain known as the state legislature, passing a law that generated millions of dollars for cutting edge research. He has a megawatt smile and the warmth to match it.

and read . . . and we looked at each other like, "Oh my God, what is this? Is this English? What IS this?" It became very apparent that I would never be the scientist who would find the cure.

What would you do if you had the winning lottery numbers in front of you, but it was very hard to get to the place where you turn them in? Would you turn your back and say, "That's too hard?" Or would you go through hell and high water to find a way? I believe we have the winning lottery numbers, within reach.

You have to stand up, and you have to be willing to fail. I've failed three times now, on passing my law in California. But I've passed it three times successfully. We've given away $15 million, and we've leveraged on another $89 million on top of that. We're over a hundred million dollars because we asked, because we stood up, because we agitated.

* * *

ADVOCACY

Dennis Tesolat

Unite2FightParalysis. Their name hit me like a sledgehammer, right in the head when I was still in the hospital, which is when I first saw their website. Just the fact that there was a group that stated that you could FIGHT PARALYSIS really appealed to me. It took me longer to think about what I could do, but the name planted the idea in my brain.

Think of it. While everyone in the hospital was getting me ready for life in the chair, when no one around me contemplated the fact that I might one day actually walk again, there was a group that was shouting out to the whole world, yes you can fight paralysis. That was my inspiration. Hospital beds are lonely places, and the people at Unite2FightParalysis made mine a much less lonely place.

So, after Unite2FightParalysis . . . got me thinking about what I could do, I did the only thing that I could think of while I was still in the hospital. I started a blog.

And I picked a name that I think says, yes we can. Stem Cells and Atom Bombs! I know some people think, what's this got to do with paralysis? The premise is, if we could build an atomic bomb in three years with less knowledge at the start than we currently have about paralysis today, then paralysis could be cured. I guess a blog is pretty much something that anyone can do . . . even though we're not the most mobile community around, we can change things.

And at each step, the way that we've encouraged people to participate – and people

Dennis Tesolat is a teacher and the general secretary of a labor union in Osaka, Japan, where he's lived for the last seventeen years. In August of 2009, a blood vessel ruptured suddenly in his spinal cord, giving him an injury at T-11. He runs a blog called Stem Cells and Atom Bombs, where his focus has been to find ways to connect people in the SCI community with short-term campaigns.

are participating — is to give people a way to help, and they will. Give people a way to help, and they will.

* * *

A couple of years ago, the Unite2FightParalysis team decided to create an annual award that would highlight and honor the efforts of the community. They called it the *Kick Ass Advocate Award*. It's a pyramid shape about six inches high, made of solid glass with the winner's name etched on one of the faces. Highly coveted! The 2010 winner was Karen Miner, who founded an organization called Research for Cure that has no offices, no staff, and no budget. Every dime she raises in her many fundraisers goes directly into the budget at the Reeve Irvine Research Center. She's been a C4 quadriplegic for twenty years.

In 2011 the winner was Geoff Kent, of SCI SUCKS fame, with his marathons and restless energy for a cure. A few months after the 2012 Working2Walk, he funded the purchase of a high-content screening microscope for Murray Blackmore's lab — a gift meant to speed up the process of identifying which genes in our DNA are blocking axon regeneration. This will change the timeframe for sorting genes from years to a matter of weeks, and all it cost was

$90,000. That's $90,000 raised by Geoff and his friends with nothing but sweat and grit.

This year, Marilyn says, the choice was obvious. The award is a double-award for two very kickass advocates: the father and son team of Don and Roman Reed. We just heard Roman deliver a stellar talk; his dad couldn't be here today, but I know that if he were everybody would be able to see where Roman gets his gift for language. The Reed team just knocks me out. They never, ever give up on the effort to find public funding for the cure, which is no doubt the reason they've succeeded so spectacularly. The Roman Reed Spinal Cord Injury Act was a simple idea: to add $3 to every moving violation ticket, collect that money, and use it to fund research for a cure for spinal cord injury. Over time, as the economy waxed and waned, that strategy changed and changed again, but during a ten-year period, the Reed Foundation saw $14 million for cure research invested directly into California labs.

The result? 175 published papers, a couple of breakthroughs, another $60 million spent in California through matching grants and out-of-state funding. As Roman said earlier, one day of trying to read the textbooks made it obvious he wasn't going to be the scientist who found the cure. Instead he and Don used their gifts to patiently, relentlessly ask that the state find a way to do

the right thing by its hundreds of thousands of paralyzed citizens. There was great celebration in the fall of 2012, just weeks before the conference, when they succeeded in persuading the California legislature to pass legislation to keep the fund alive, this time with just a $1 fee on traffic tickets. Then, a blow that would have crushed any normal person's spirit.

Governor Jerry Brown vetoed the bill on principle, saying that he doesn't want a fee-based economy. It's not a fair way to fund public projects. And yet there was no money in the state's General Fund to keep the Reed Act going – that was the reason they went to a traffic ticket add-on in the first place. Don Reed's public reaction? *"I do not hold a grudge against Jerry Brown. He had an impossible task to perform."* Also: *"We'll be back."*

He wasn't kidding. By February 2013, there was already another version of the Roman Reed Spinal Cord Injury Research law making its way through the California legislature. This time, Don and Roman Reed have asked that $2 million from the state's General Fund be appropriated for use in California research labs. Whatever happens to it, I would never bet against Don and Roman. They're not giving up, as Roman said today, until everybody's walking. That's kickass.

* * *

The full conference is about to split breakout sessions, which is always one of the tough moments for me at Working2Walk. How do I decide which small group to join? There will be three consecutive forty-minute sessions, and during each of those there will be four options. Something has to give.

chapterTen

Alex Aimetti: Businessmen in Our Corner

The Langer Lab is on the front lines of turning discoveries made in the lab into a range of drugs and drug delivery systems. Without this kind of technology transfer, the thinking goes, scientific discoveries might well sit on the shelf.

New York Times, November 24, 2012

It's mid-afternoon, and the flood of information has left some of us wanting to just push through the doors to the spacious hotel patio and doze in the sun near the potted palm trees. As my friend Jennifer Longdon put it, this conference can be like trying to drink from a firehose. For those with the stamina, one option is to stay right in the ballroom and listen to Alex Aimetti, who's a manager of research and development at a company called *InVivo*; he's here to describe his company's current agenda and progress. The talk will have a different flavor from what's gone before, because InVivo is frankly and happily planning to make a *lot* of money curing paralysis. (More power to them, I say.)

Alex, it turns out, is a young chemical engineer who's only been working at InVivo for about seven months. He came there from MIT's famous Langer Lab, which is a sort of giant breeding ground for new product patents (811 of them) and start-up companies (25 so far) in the field of biochemistry. One of those companies is InVivo. The short version of his presentation goes like this: *InVivo wants to be the company that builds state-of-the-art intelligent tools for delivering combinations of therapies into damaged spinal cords.*

What does that mean? Think about the first time you heard about those stitches that don't have to be removed because they just dissolve after a wound has healed. Those are an early example of a bio-engineered solution. Chemists can make them tough and long-lasting for really big wounds, or extra stretchy for wounds in places like knees that need extra freedom of movement. They can engineer them, in other words, in such a way that they're perfectly designed to mesh with whatever body part they're placed in. The same thinking has produced biodegradable screws that make repairing facial bones much simpler and less painful than it used to be. There are sutures, stents, and gels – a whole array of tools that added up to a market valued at $26 billion in 2008 and is projected to be worth $65 billion by 2015. This is the prize InVivo wants to win.

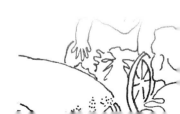

So, what would chemically engineered products for spinal cord repair look like? One possibility is a *polymeric scaffold* that can be made very soft and then shaped to the exact configuration of the lesion in a damaged cord. *Polymers* are compounds made of millions of repeating molecules, all chained together. Silk, rubber, amber, and wool are all naturally occurring polymers; nylon, vinyl, and silicone are familiar examples of synthetic ones.

InVivo's polymeric scaffold has been extensively tested in spinal-cord-injured rats and monkeys, with impressive results in acute situations. Alex shows us a video of a pair of monkeys, both of which had thoracic injuries (hemisections). The one that got a polymeric scaffold is much more mobile that the one that didn't . . . I keep waiting for him to show us the data, but he never does. The papers that InVivo has published (five of them between 2002 and 2010) about their scaffolding technology are all available at their website, though, and I've put a link to it in the bibliography.

Today what Alex wants us to know is that the scaffold *all by itself* seems to be a pretty good therapy for new injuries, even without adding any of the cells, growth factors, enzymes, or gene knock-outs the rest of the scientists are developing. It acts to spare healthy tissue, and InVivo expects to be given the go-ahead to try the scaffold in human patients sometime in 2013. This is just one place where the Langer Lab expertise matters so much. Bob Langer sat on the FDA's Science Advisory board for seven years; his team knows the system from the inside and gets how to work within it to move things along quickly. He's also on the scientific advisory board at InVivo.

One factor that will speed things up is that the polymer scaffold is – at first – not going to be carrying any drugs or cells into the human subjects. It's just a device, and one that's mostly made of materials already approved for human use. The process for getting new devices approved is much simpler and quicker than the one that brings new drugs to market. Once InVivo has shown that it's safe to put the scaffold into an injury site, they'll be positioned to get permission to engineer combination therapies.

The other new product is a smart *hydrogel*, and it's being designed to work much like the scaffold does – as both structural support inside the damaged cord and as a potential delivery system for all the kinds of therapies now being tested in animals around the world. Like the scaffold, InVivo's hydrogel is made of polymers. There are hydrogel-based products all around us, including disposable diapers, burn-healing and skin protection gels, breast implants, contact lenses, and soil correctives engineered to hold moisture.

Alex recites a litany of the kinds of questions it's his job to answer with

Why do we have an FDA? In 1937 a raspberry-flavored, off-the-shelf product called Elixir of Sulfanilimide that hadn't been tested at all killed more than a hundred people in a matter of weeks. The company that made the elixir paid no penalty, because they hadn't broken any laws. The owner said, "I do not feel that there was any responsibility on our part." Congress created the FDA the following year; all products would have to pass safety standards in the future.

respect to this material:

> *So as I lead the program in developing the hydrogel technology, there are a lot of questions that I need to think about and answer . . . how long do I want this material to be around? Do we want it to be around for days, weeks, months? We can tune that. That's one of the things that's very versatile about these biomaterials. What type of mechanical properties are we looking for? Do we want something more rigid? Less rigid? What type of therapeutic release profile are we looking for? A lot of these discussions come to the great researchers around here that are identifying new molecules. What type of delivery regimen are you using now? Are you giving it on a daily basis? Why do you need to do that? What type of therapeutic level are you trying to hit on a day-to-day basis? And we then try to achieve that release profile, using a biomaterial with just one injection.*

And now at last we get to talking about chronic injury – the whole reason for Unite2FightParalysis' existence and for this conference. So far, what we've heard is that Alex is running a research and development team that's working closely with surgeons and representatives of the FDA to build and test polymer scaffolds in people with new injuries. New injuries, though, are not going to represent much of that $65 billion market we heard about a few minutes ago. InVivo's ultimate plan is to heal the millions of people who are living with paralysis now, not just those who get injured in the future.

What gives me a good feeling is recognizing that InVivo has made ten years' worth of serious investment in this project, and that they fully expect to succeed and make fat returns on that money. Alex describes their new corporate offices and labs in Cambridge, MA, where they maintain a "vivarium" with 400 rodents, all for testing spinal cord injury therapies. They're staffed with chemical engineers, biomedical engineers, mechanical engineers, and neurosurgeons. They have what's called a "GMP" facility to produce their scaffolds and hydrogels, which means that the materials coming out their doors will have been made in such a way that they meet federal standards for safety and purity.

> The 1938 law that followed the debacle with poisonous raspberry-flavored elixir included provisions for GMP, or Good Manufacturing Processes. It requires, among other things, meticulous record-keeping, control of operations, training, and ability to recall products.

So. How does all this add up to a cure for chronic paralysis? The idea is that the scaffolds and hydrogels will be seeded with whatever combinations our friends in the neuroscience community can prove are effective. A hydrogel might come programmed, for example, to first release some peptides, which Jerry Silver just showed us acts to neutralize the glial scar. Then it could deliver a shot of growth factor so that the axons – newly freed to move past the injury site – would do so quickly.

Or how about a scaffold that contained neural stem cells in combination with the knockouts for PTEN and all the other genes that prevent neurons from

growing? The possibilities really are endless. InVivo is aggressively positioning itself to be ready with FDA-approved and human-tested devices, so that when the moment arrives to try those first combinations in chronically injured patients, getting them into people's cords will be easy.

* * *

Coda: As I write this in April, 2013, InVivo has already hit a giant milestone. On the fifth of this month, the FDA gave them the go-ahead to start testing their polymer scaffold in human patients with new spinal cord injuries. By the time we all get to Working2Walk 2013 in Boston, there will probably be patients with Alex's scaffolds in their cords. Here's what InVivo CEO Frank Reynolds said in announcing the news: "... *for the patients and families . . . every day can feel like there is no way out. We expect a successful safety study to provide not only the first treatment for acute spinal cord injury, but also a safe platform for next generation treatment options.*"

Frank Reynolds has a spinal cord injury himself; he's not going to be satisfied until he gets a product for chronics to market.

chapter**Eleven**

Leif Havton: Cauda Equina Injuries*

> *California's stem cell agency closely monitors progress in stem cell research and targets funding in areas most likely to lead to new therapies. We're funding great research . . .*
>
> CIRM Website

After a much-needed break, a small group of us gather back in the main ballroom to look at one last set of slides. The main program is over but there are small breakout gatherings getting underway, and we're about to get a chance to hear Leif Havton speak. He makes me think of my childhood in northern Minnesota even before he starts to talk; he has that broad, reassuringly kind face that I associate with the Swedish immigrants who were my teachers and neighbors a long time ago. (He sounds like them, too.)

---

* Author's note: Because this presentation wasn't part of the main program, I'm including it in this chapter with the other breakout sessions. It actually took place at the very end of the second day.

Leif holds a Ph.D. in human anatomy and an M.D., both from Umea University in northern Sweden. He runs a research lab at the UC Irvine's Department of Anesthesiology and Perioperative Care, where he focuses on nerve root and *lumbosacral* injuries.

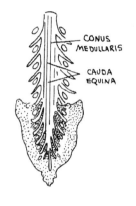

Very few labs are working on strategies for those injuries; they're a sort of red-headed stepchild in the SCI research world, partly because they only represent about a fifth of all injuries, and partly because the mechanism that causes impairment is different. You could argue that the very lowest ones aren't even spinal cord injuries, though people who have them are certainly paralyzed. We need a little anatomy to understand why. The spinal cord starts at the base of the brain and ends at what's called the conus medullaris, which is, roughly, Latin for cone-shaped thing at the marrow. That cone has the shape of a stalactite or a bullet, long and narrow with a pointy tip. It's at the level of the first two of the five lumbar vertebrae, L-1 and L-2.

Spraying off the cone is a tangle of downward strands that look a little like the tail of a horse, which is why it's called *cauda equina*, a phrase that is – naturally – Latin for *horse's tail*. The horse's tail goes down from the cone to the bottom of the sacral segments. That tail is not part of the spinal cord – it's a collection of spinal nerves. Each nerve is a bundle of axons that go directly to the pelvic organs and muscles in the legs and feet. And what that means, as Leif is saying, is that in lumbosacral injuries you get a direct hit to the neurons that innervate the bowel and bladder, as well as to the motor neurons that attach to muscle fibers in your hips, thighs, ankles, and so on.

I think of the big difference between these injuries and the ones that happen higher up this way: the spinal cord is like a massively complex superhighway, and the spinal nerves are like exit ramp bridges that split away to distant destinations. In complete cervical and thoracic injuries, there's a crater in the middle of the highway, and sometimes the bridges nearby are damaged. In complete conus and caudal equina injuries, the bridges have been blown up. There's information coming all the way down the cord, but it stalls at the junction between the central nervous system and the peripheral one – the junction that's formed by spinal nerves.

This is why people with complete cauda equina injuries don't have spasticity; there can't be any signals at all to their legs and feet because the spinal nerves that usually carry those signals are gone. Their lower limbs are always in a rag doll state. Unfortunately, and for reasons not clearly understood, they often have considerable pain. In higher injuries, garbled signals come through

the damaged cord and make their way to the legs through intact spinal nerves. It's those mixed-up, useless signals that cause tone and spasticity.

How do you go about fixing damage to spinal nerves? In the fall of 2010 – two years before this lecture – a $1.6 million grant allowed Leif and his collaborators to develop a cell-based therapy. Here's another difference between lumbosacral and higher injuries: damage to motor axons inside the upper cord doesn't do anything to the cell bodies up in the brain. But damage to axons in the spinal nerve *does* cause damage to their motor neuron cell bodies. In fact, it kills those neurons dead, and the more time that passes, the more neurons die. Leif says that this makes a lumbosacral injury sort of like stroke or heart attack, where it really matters that you get treatment quickly. It also means that those lost motor neurons are going to have to be replaced.

The talk today isn't going to be about that, though. He isn't going to tell us about the work he's doing to replace the dead motor neurons, because that project is still in progress. That project involves growing motor neurons from embryonic stem cells (as Hans Keirstead has already done) and transplanting them into the damaged conus and cauda equina. If all goes well, we'll get to hear about it next year.

* * *

What we're going to hear about today is a strategy he's worked out over the last ten years or so to repair the broken bridges quickly by executing a repair surgery. The surgery strategy is for acute injuries, and Leif walks us through a detailed history of how they developed it and how it works. To begin with, you need to create an animal model that closely represents what happens in people. (Sorry, rats.) They did this by tearing out (*avulsed* is the medical term) the *ventral* nerve roots at the surface. *Ventral* means front; Leif was destroying the bridge that carries motor neuron axons out to the muscles and the bowel and bladder.

Next they reversed the process by surgically reconnecting some of those torn out nerve roots. They had reason to believe this might work; a similar surgery had been done at the Karolinska Institute in Sweden, where a doctor named Thomas Karlstead had reconnected nerve roots that exit the cord way up near the shoulder. Why not try the same thing with lumbosacral nerve roots?

When they did that surgery, what they saw was a doubling in the number of surviving neurons. Leif published this work in 2006, and he shows us some data from that paper. Not only were lots of axons growing, but Schwann cells arrived to myelinate them, which means they'd succeeded in growing into the periphery – and that was of course the whole point. The bridge had been repaired, at least in part. When they put slides of these repaired nerve roots

under an electron microscope, they saw that the Schwann cells were not only still alive, they were multiplying. They were secreting growth factors. They're the reason, probably, that those axons were growing out of the surgically reconnected nerve roots. To axons, Schwann cells are like bread crumbs.

Next was a collaboration with a couple of Johns Hopkins scientists who helped them put growth factors inside biodegradable nerve guidance tubes, which only convinced them further that they really were able to use a nerve-reconnection surgery to get axons to grow into the periphery after the ripped out ventral root event. Still, they had not seen a return of function.

Leif's team expanded their injury model. They avulsed spinal nerves at multiple levels on both sides and surgically replanted on both sides. The goal was to show that they could restore bladder control. How do you measure this? (Everything has to be measured in science, or it doesn't count. Science comes from an ancient word that means *to know*. If you can't measure it somehow, the thinking goes, you don't really know it.) Leif goes into some detail about how they went about measuring restored bladder function. It involved a pressure-recording catheter and some electrodes on the sphincter; pretty much the same bladder evaluation procedure that you or I would get at the urologist. A rat with destroyed ventral lumbosacral nerve roots can't pee no matter how much saline they pump into its bladder, because there aren't any axons around to fire up the right muscles. But the rats with their nerve roots surgically repaired did get back that reflex, about twelve weeks after the surgery.

Another area they tried to address was pain. Because the repair surgery inflicts a new injury to the cord, you have to be concerned that you're going to cause pain. They did another animal model, but this time they only removed motor fibers and left the sensory fibers intact. The goal this time was to make sure that replanting those motor fibers didn't cause that nasty light-touch pain known as *allodynia* – the feeling that makes you want to jump out of your skin when sheets brush against your bare legs.

What they found was that even though they hadn't removed any sensory fibers, the rats still got allodynia . . . and then that pain gradually disappeared after the replanting surgery. This is unexpected, right? Why should you get neuropathic and visceral pain without any damage to the sensory neurons? And why should it be made better by surgically replanting the motor nerve roots? They found out later that the Swedish doctor who was doing replanting of the shoulder nerves had learned from his patients that their pain also gradually went away after surgery. Leif's team had inadvertently demonstrated with rats that something similar was at work.

He ends with a summary of what makes this acute surgery an appealing strategy for acute lumbosacral patients. In the animal models, it kept a lot of motor neurons alive that might have been lost. It allowed axons to grow through the grafts and out to the muscles. It improved bladder function. It lessened injury-associated pain. And best of all, it's just another surgery. The FDA doesn't have to approve it, and no one will need to get permission to test it. The icing on the cake is that it could be done while the usual decompression/stabilization surgery is happening.

*   *   *

Fair enough. The often over-looked cauda equinal spinal cord injury has a champion. He's got a procedure that has been shown to help acute patients, and he's got a plan to develop a stem-cell-based approach that will help chronics. It's time (again) to be glad that the voters of California chose to invest in cures, because if this treatment is effective, the research that moved it forward will have happened on their dime.

chapter**Twelve**

Up Close: Os, Jerry, and Murray

> *I think that the biggest hurdle is actually before the FDA. Say I make a discovery, and I do the experiment, and I publish that result. But then I have to go back and really work out the details. And it's really the details that take a lot of experiments. What's the right time to give a drug? How many times do you need to give it? How long do you need to give it? What's the dose that's right? Each of those is a separate experiment, and that's the preclinical work that actually takes a long time to get to, even to begin to get to the point where you could get to the FDA. It's that preclinical work that's the hardest to actually get through, and mainly because of timing.*
>
> Os Steward

My last stops today will be in a couple of classrooms where this morning's presenters are taking turns answering questions. Right now Os is up front, leaning against a table with his arms folded. He's casual, facing twenty or twenty-five of us. The schedule says the next hour or so is Os' gig, but Jerry Silver is here, too. At some point he'll take over as the main

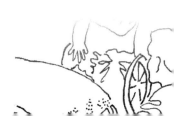

responder. (Note: this reads like a transcript, but it's not. There was no taping going on during these breakouts, which means that what follows is my best attempt to reconstruct my real-time notes into something useful.)

Q: Why is the research so slow? Is it lack of collaboration?

Os: *That's a common perception . . . in the spinal cord injury field, I really believe that we're very collaborative. The problem isn't that we don't want to do it, it's finding out how to do it. Sharing data isn't collaborating. I've been blessed with private funding that allows me to move people off other projects, which I can't do if I'm paying people on a grant. The bottleneck is the funding. NIH does have some multi-investigator grants, but they're hesitant about giving great big sums all at once.*

Q: What about competition — don't you compete for grants?

Os: *We do . . . it used to be that you could get funded even if you were in the middle of the pack. Now people are trying to choose the best 8 or 9% of all the applicants.*

The PTEN work requires us to go next into other animal models . . . but that means these won't be innovative studies, they're not ground-breaking. That kind of project doesn't get funded, because everyone at NIH is looking for the next big idea. This is one of several valleys of death.

Q: If we were to start aiming our energy around the Manhattan

Project idea, what would that look like? What would those steps be?

Os: I threw that out as a challenge . . . but the fact is that if someone said they had a billion dollars, it would happen. I directed the way we spent the Roman Reed funds; I'd just established the Reeve-Irvine Research Center, and the state gave the funds to the university, who gave them to us. I called a town hall meeting. We sat in open session all day long and at the end of the day came up with a plan.

The point of that story is that people are perfectly able to come together to make something happen. A Manhattan Project for spinal cord injury would take a lot of time – obviously more than a one-day town hall, and a lot of very careful planning. And if I were to take that on (directing it, he means), *I'd have to fire everybody in my lab because I wouldn't have any money to pay them.*

Think about the California Institute of Regenerative Medicine approach, which is to see the process as a kind of long train, with engines needed for each of the cars. The engines are the individual labs with their individual skill sets. I have a lot of expertise in the basic science part of things, but not so much in the next phase. The Institute breaks up the money so that each car is being pulled efficiently by the engine that has the most power for that task . . . so to speak.

In my lab, fifty thousand dollars is huge. That's one more person working for a year to get more accomplished. We rejoice over that kind of money. And one hundred thousand is

somebody capable of managing a couple of people. Again, huge.

Q: Why doesn't your gene therapy cause cancer? It seems like turning off a gene that tells axons not to grow would have to cause uncontrollable growth.

Os: It's not a cancer worry because cancers don't arise from neurons. (Having dealt with that question, he goes on a riff about getting the gene therapy off the ground.)

We also know that no matter how much you boost up the ability of nerve cells to grow, they're not going to get past the cavity that exists around the lesion. Tomorrow Mark Tuszynski is going to talk about using a certain kind of stem cells to fill that void. Having said that, the problem becomes how to test each of these separately in an FDA-compliant way. Suppose we wanted to knock down both PTEN and SOCS3. We could do it, but then we'd still have to add Mark's stem cells, and that would be another set of FDA tests.

That's not necessarily as big a burden as it sounds. Think about how the FDA actually works — you start with a pre-IND meeting, in which you talk about what will need to happen and plan. (Meaning, it's not like they just force you to keep doing things over and over because they don't get what you're trying to accomplish. IND stands for Investigational New Drug. In pre-IND meetings, you start by explaining what you plan to accomplish, and they work with you to set up an efficient path to get there. What Os is saying is that if he gets to combining his gene knockouts with Mark's stem cells, it will be done in a carefully staged way, with full communication with the FDA throughout the process.)

Q: Are you targeting acute or chronic?

Os: We hope that what we're working on will work for both. As Bob Yant said earlier, it's faster to work with acutes to find out what you want to know.

Q: It seems like mathematical craziness to do human trials with acutes, because you don't have a population to work with.

Os: Okay, let's say you only found an effect in acute models, like in the Geron trials. The transplants into acutes worked, the ones into chronics didn't. (Geron got FDA permission to test its cells only on people with brand new injuries; this wasn't because they only wanted to help those people or because it was most efficient. It was because in the lab, the cells worked on animals with acute injuries – but when they tested them on animals with chronic injuries, nothing happened.)

At this point the time allotted to Os is about up. Someone is arguing that we should always be doing chronic studies. It's time for Jerry Silver to take questions, but first he takes a minute to chime in on that subject. He says we should just spend the money and work on chronics, then goes into a little speech.

Jerry: If you believe in your science, if you think you've got something that works, why not be focused on working out the problems that the chronic cord throws at you — namely, the big scar. I say, with respect to chronic or acute, just do the chronic. Every lab that has a robust result, they should be doing it. Also, at NIH they're dying to see work related to chronic injuries.

Think about what's going on inside that chronic cord: axons that were cut have died back. There's a dense wall full of stem cells surrounding the lesion with a moat full of nothing in between. Those axons can sit there for decades. That very dense scar has to be removed, and then you can take advantage of plasticity with focused rehab – like locomotor training.

What's going on in the cord is both good and bad. Good: the nerves are still there. Bad: things are getting reorganized in nonfunctional ways. But that abnormal circuitry can be fixed with lots of therapy. In the case of the respiratory system, just chondroitinase (ChABC) restores proper function. What happens if we just knock out that PTEN? We might get chaotic new circuitry; that's one of the things we're testing to see. We're doing that right now. In chronics. We have a whole variety of lesion models, from hemisections to removal of a whole section of the cord to contusions. Contusions are the bloodiest, and therefore the biggest.

Q: What kind of animal would recover the fastest with a transection?

Jerry: No warm-blooded species can regenerate any function with a complete transection, but an embryo does have more plasticity and can recover better from a partial transection.

Q: What do you think is going on with those animals that didn't recover in your lab?

Jerry: All of our work is double-blinded, and we've done the work three separate times, so I know it's not a fluke. We try to do all the injuries exactly the same. Then each gets the exact doses that worked well in tissue culture — and still we saw the variability you heard about this morning. Some animals got a lot of walking back, some got none at all. Some got urination, some got none. Some got both, some got neither. Why? (Shrugs) Maybe the injuries are not exactly the same.

Q: Who are the young guns that are going to replace you?

Jerry: We're always training new students; several from Os' lab are right here. Also I'm not dead yet!

Q: What's next for your peptide therapy?

Jerry: What you've got to have in order to go forward is a robust result. You don't need a little tick in a graph. You need something big. The peptide work had a robust result; in fact my university got all excited about this and wrote a patent, which they're now shopping around. We're in our fourth round of discussions with GlaxoSmithKline, though they want it

for stroke and other things. Once they get it, we lose control. They can do whatever they want, and spinal cord injury is not going to be high on their list. I won't sign any contract without something in it for spinal cord injury patients. This community can influence that process, but only if you make a lot of noise.

Participant comment to the room: Why do we rely on pharmaceutical companies at all? They're not there to help us. We need to be like the AIDS community was — we can raise money ourselves, independently. Pharmaceutical companies just want to make Viagra 4.0 or something, a better boner medicine.

Jerry (and all the rest of us): laughter.

Q: Is there anything we could do ourselves to get rid of the scar?

Jerry: No.

Q: Could you talk about rehab a little?

Jerry: What Mother Nature is doing during rehab is trying to stop chaos from developing. One of the earliest populations of patients who ought to be tested are those patients who have worked very hard in rehab and reached a plateau. Just a little boost, according to Susan Harkema, might change the outcomes.

* — *

Okay. It's four in the afternoon now, and we've been at this for eight straight hours. I check my schedule; what's next? The last event of the day will be one more question-and-answer session, this time with Murray Blackmore. I pack up my laptop and head off to find his room.

There are only six of us at the start, and Murray looks like he might be getting as tired as I am. The session starts off with a discussion among a few participants about one of the central problems of advocacy: finding the balance between acceptance of your situation and pushing for a cure. One guy says that he thinks his friends get hung up on the need to seem macho-like, they try to do that by being great at sports, which then becomes a sink for their energy and time. It's a great way of reclaiming their health and energy. It's good – and yet it's also turning away from the push for a cure. Murray sits quietly, listening. We don't seem to be able to come up with any good questions for him.

I noodle around online, looking for information about him. He has a lab at Marquette University, which is interesting to me. Marquette University is a fine Jesuit school in southeastern Wisconsin, but it doesn't seem like a likely landing place for someone who trained at Stanford and worked at the Miami Project. The Miami Project is something of a mecca for people who want to

> You have to admire the guy! This is what he once wrote for a bio:
>
> I attended Stanford University as an undergraduate, where I conducted research on the impact of fire and cattle grazing on dry tropical forests. I spent my summers in Hawaii doing field work, which was not unpleasant. After graduating I spent a year teaching fourth grade as an Americorps volunteer, and then spent a year bicycling from St. Paul, Minnesota to Santiago, Chile. I always knew I was destined for research.

cure paralysis. It has fantastic facilities and hundreds of researchers, all focused on just one thing: spinal cord injury.

Murray went from Miami to this school in Wisconsin where the focus is on something completely different: ending cocaine addiction. But he's still doing spinal cord injury work. Why the move? I don't get to ask him this time, because he's saying something interesting about being in a place where scientists are focused on a completely different problem.

Murray: Some new ideas — completely new ideas — are there in front of me. I think that if there's to be a Big Project sometime for spinal cord injury, it would need to be seeded somehow from outside. The sparks are seeming to come from outside the big centers for spinal cord injury, like the Miami Project or even the Reeve Irvine Research Center. The PTEN discovery started at Harvard. You still need the big SCI centers, though, because there has to be someone who recognizes the potential of that spark.

* * *

He's talking about how hard it's going to be to test all the combinations of genes that might be involved in preventing axons from growing:

Murray: Right now it takes between four and six months for the Miami Project to grind through seven hundred genes. The problem is that not many will have an effect all alone, so you need to test for combinations—which just explodes the number of things you must test. *Another thing — screening is really expensive. I've been through about a thousand genes, and twelve showed promise in the culture dish, and none of those in the animals.*

It actually took many molecular tricks to get just one to really work. The goal would be to work with your most likely 200 candidates and inject them in combination pools in animals, because it wouldn't make sense to waste time with dish experiments when so many fail anyway. That would be expensive, but it would get the job done.

> What does he mean about the number of things needing to be tested exploding? If you have 50 things to test one at a time, that's 50 tests. If you want to test them in all the possible pairs, there's a formula (called the binomial coefficient) to figure out how many tests that would be. The answer is 1,225. Big difference.

* * *

Made it! That's the end of the first full day at Working2Walk 2012. We're about to go out on the patio and enjoy some lovely California food and wine. *Halleluiah.*

Bench to Bedside

chapter**Thirteen**

Jonathan Thomas: Follow the Money

> *Oct. 22 2004-- Proponents of Yes on Proposition 71, the California Stem Cell Research and Cures Initiative, today announced the launch of a new television ad featuring Christopher Reeve, who was paralyzed in an equestrian competition in 1995. Mr. Reeve passed away on October 10, 2004 from heart failure. Mr. Reeve not only put a human face on spinal cord injury but he motivated neuroscientists around the world*
>
> PRI Newswire

Working2Walk Day Two begins with what's billed as a "Keynote Speech," so named because it's supposed to set a tone for what follows, sort of like the little note that one member of a barbershop quartet plays on a tiny harmonica just before the harmony starts. *A key note.* You know, sometimes I forget to be amazed. Here's this crazy little group of people who want there to be a cure for spinal cord injury. Christopher Reeve dies suddenly, and they decide to have a rally because they don't know what else to do. The

rally turns into a science and advocacy symposium. The science and advocacy symposium becomes an annual thing. They're still just a crazy little group of people, though, and now here before me is a tall, long-faced man named Jonathan Thomas: J.D., Ph.D. He's a very big deal – the chair of California's famous Institute for Regenerative Medicine, fondly known as CIRM. I forget to be amazed.

Some background is needed, and it can't be told without a bit of history and politics. Bear with me. In November 1998, toward the end of Bill Clinton's second term as president, a team of scientists working out of a lab at the University of Wisconsin in Madison figured out how to isolate and grow batches of human embryonic stem cells. These are the cells that – given the right cues and conditions – can either make copies of themselves forever or turn into any of the other kinds of cells in the human body. Every one of us once existed as a batch of these cells, and now here they were in a laboratory dish, endlessly replicating, waiting to be given their instructions. *Become part of an eye. Grow into muscle.* President Clinton organized a task force to consider the ethical implications of what was a momentous discovery, and that group delivered their findings and recommendations a year later, just before he left office.

One of the questions the task force tried to answer was: Would it be ethical to do medical research using these cells? Under what circumstances would it *not* be ethical? Also, should American taxpayers help to pay for it, as they routinely pay for other kinds of medical research? The task force recommended that the National Institutes of Health (NIH) fund such research, but only under a set of guidelines they had carefully outlined. Taxpayers should pay for it. The source of the cells, though, was problematic to many people – human zygotes that were being kept in a frozen state by the millions in fertility clinic storerooms.

When infertile couples use a common medical procedure (known as in vitro fertilization, or IVF) to fertilize eggs outside the woman's body and later implant them, there are almost always a small number of fertilized eggs "left over." It is these frozen embryos, a few days into the human life cycle, that are the source of the precious stem cells. Under the Clinton guidelines, donating them would work by the same kind of process that people use to donate their organs to science, with full permission and no money changing hands. Most of the frozen embryos were destined for the medical waste containers; their only purpose was to be a sort of backup, in case the embryos that had been implanted didn't produce a successful pregnancy.

George W. Bush was inaugurated in January 2001, and within weeks

> For many people, it was the idea that embryos could be thrown away but not used to discover cures that rankled. Like most Americans, I have friends who used the IVF process to start their families. One of them told me that when their twin daughters were about twelve, she and her husband asked for the remains after their long-frozen leftover embryos were destroyed. They wanted to somehow mark the event; the clinic told them that was impossible, due to rules about medical waste products. The embryos were thawed and disposed of.

he'd put a hold on any planned research funding and asked for a review of this policy; he wanted to weigh in himself on the idea of using these embryos as a source for cells to be used in research. Late the following summer, he delivered what would become the new federal policy. There were already a small number of "lines" of human embryonic stem cells in existence. These —and only these— could be used in research funded by the federal government. In the eyes of many scientists (and a lot of people eager to find out if these cells could help them), the new policy amounted to a ban.

* * *

After that, a group of patient advocates in California decided to take matters into their own hands. If the USA as a whole would not fund research in what seemed to be a very promising new area of medicine, they would ask the people of their state to do it themselves. Believing that it would be unethical *not* to find cures, they wrote what became known as Proposition 71, got it on the ballot for the 2004 election, campaigned like mad, and were rewarded by the voters with three *billion-with-a-B* dollars, all to be spent on regenerative medicine inside the state of California. They created the California Institute for Regenerative Medicine.

The first funds were released in 2006, and in the years since, California universities and companies have formed a stronghold in stem cell research and regenerative medicine that's unrivalled anywhere on the planet. It was a bold move and a big investment, and Jonathan Thomas, who oversees the spending of that money, is here to set the tone – to deliver a keynote – for today. Most of this talk, it turns out, will not be about spinal cord injury research.

Instead, we hear about continuing funding challenges in the federal government. The congress has been unable to work out a sensible process for writing a budget, and there is a possibility that across-the-board spending cuts will be enforced early in 2013. Research labs all across the country depend on federal tax dollars to keep their fragile, carefully planned experiments on track, and the cuts would mean scaling back. Competition is already fierce, and the across-the-board cuts will only make it more so.

In spite of current problems in money and politics, says Jonathan, the future of medicine is going to be personalized. Someday, each of us will be able to use the potent combination of our own genes and some custom-built cells to heal conditions that were formerly incurable. He describes the importance of genetic sequencing with a long story about Alexis and Noah Beery, a pair of California teenagers who went from being profoundly disabled to being athletes, thanks to the mapping of the human genome. It's a great story, but it seems a

Alexis and Noah are fraternal twins whose parents both happen to have mutations on the same gene. What are the odds? The whole genome has between 20,000 and 25,000 genes. If only one of the parents had passed on the mutation, both the kids would have been fine.

bit out of place in this room. Nobody is here because of a genetic disorder. The principle, of course, does apply: knowledge leads to success in ways that couldn't be predicted.

Jonathan says that reporters often challenge him with questions about where the cures are. *CIRM has spent more than half of that three billion dollars since 2006, and where are the cures you promised?* He says he tells them to think about a reporter questioning the people who were trying to cure polio in 1955. Scientists had been funded by the March of Dimes for seventeen years at that point, and *where was the cure?* His point is that science takes time. CIRM is targeting a huge portfolio of incurable diseases and conditions: Type I Diabetes, HIV/AIDS, heart disease, macular degeneration, Alzheimer's, Lou Gehrig's disease, and yes, spinal cord injury.

Here's a partial list of CIRM's investments in spinal cord injury research to date, as shown on their website in April, 2013. I chose these people because each of them is involved in the work that's being discussed here at Working2Walk. All amounts have been rounded to the nearest million.

Stem Cells, Inc.	$20 Million
Geron Corporation	$25 Million
Zigang He	$6 Million
Mark Tuszynski	$5 Million
Leif Havton	$2 Million
Aileen Anderson	$1 Million
Brian Cummings	$1 Million
TOTAL (These grants)	**$60 Million**

Outside that list is another four grants worth an additional $5 Million. CIRM has spent $1.8 billion in the last six years, and of that, $65 million went directly to spinal cord injury projects. For those who like arithmetic, that's about thirty-six bucks out of every thousand spent – about a third of 1%. It would be discouraging, but I'm thinking about what Murray Blackmore said yesterday: *The sparks will probably come from outside the field.* It's also just a fact that heart disease, Type I diabetes, and AIDS are destroying whole populations of people. Spinal

cord injury – devastating as it is on an individual level – can't be compared to those things.

Keynote delivered, tone set, we're going to move right on to a pair of talks that will follow the trail from basic science to animal testing to human trials – from bench to bedside. Aileen Anderson, Brian Cummings, and StemCells, Inc. will all be part of this story that the California Institute for Regenerative Medicine is underwriting. *Thank you, California taxpayers.*

chapter**Fourteen**

Aileen Anderson: Start with Good Science

It's very tough for patients and their families. We believe stem cell therapies could provide significant functional recovery, improve quality of life and reduce the cost of care for those with spinal cord injury. That's our goal.

Aileen Anderson

She's *focused*. That's the word that comes to mind when Aileen Anderson speaks. Her face looks almost carved, and the clean lines all seem to point toward whatever she's saying. It's impossible to imagine her lapsing into sloppy thinking or speech, which is remarkable when you notice that she's talking at something like twice the ordinary rate of conversation. This talk is going to take us back to experiments and data, but today is not going to be about genes, molecules or growth factors. Today is going to be about cells.

She starts with a brisk review of how the cells in our bodies branch off during development, using a watercolor-style image of a tree to illustrate. There is the trunk, which represents the kinds of stem cells present in the earliest embryos. The trunk branches four ways, and one of those major branches becomes the family of cells called the *ectoderm*. The ectoderm branch splits again into smaller branches, one of which symbolizes neural cells, the ones destined to form the nervous system. Other branches on the ectoderm side of the tree are sense organ cells, which will become eyes and ears, and dermal cells, which will become skin and hair.

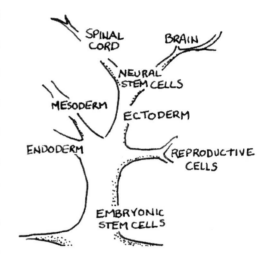

Why does this matter? Because one cell-based therapy approach is to replace lost cells, or to add cells that will interact with surviving cells in a way that leads to recovery. Cells are not just structure, like the fibrin that Os Steward talked about yesterday. They're little organic machines that do specific jobs. And if you're going to let someone put cells – living things with their own agenda – inside your body, you want to know exactly what those things are going to do.

In Aileen's lab the cells they work with originate in the place where the ectoderm branch splits to form the neural branch. Those cells are called *neural stem cells,* because while they will definitely be cells in the central nervous system, it's not yet determined which of the three kinds of central nervous system cells they'll be. They could become *neurons*, which carry the messages, or *oligodendrocytes*, which wrap the axons in insulation, or *astrocytes*, which I like to think of as utility players.

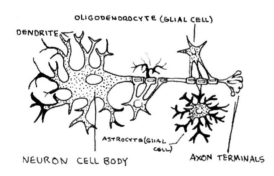

Not very long ago, scientists thought of astrocytes as a kind of filler, as if they had no function but to form the tissue of the brain and cord. It turns out that they're all over the place in terms of pitching in: they regulate signal transmission, they provide fuel and nutrients to neurons, they stimulate oligodendrocytes to make myelin, and that's just the beginning.

Anyway, in the sense that neural stem cells can still become any of those three cell types, they're unformed. What Aileen's team does in her lab is test to see what happens when human neural stem cells are put into rats and mice with spinal cord injuries. What would they become? Would they stay alive at all? She points out that only twenty years ago, nobody had a clue that human neural stem cells are still functioning in our bodies – and that the science around them has been moving forward at rocket speed.

Nature, nurture, niche. This is how Aileen describes the three factors that will determine whether a treatment is effective or not. *Nature* is about the thing she just explained – where is the cell along that developmental tree? What is its nature right at the moment it's injected into the spinal cord? *Nurture* has to do with what happens to the cell as it's being prepared for that injection. What needs to be done to those cells in a dish in order to make them most productive? Finally, *niche* refers to the little world of the injury site, with the scar and the cavity and the trapped axons we heard all about yesterday.

Now this is interesting. The neural stem cells that her lab has been using are what she calls "tissue-educated," which means that they grew in living tissue and not in a dish. The holy grail of stem cell scientists is to figure out how to take one of your own cells – say, a skin cell – and turn back its clock so that it moves backward down that developmental tree. This is called *induced pluripotency*, and there's a reason it would be a great solution. Your cells all have your DNA, and your immune system wouldn't go after them the way it would rush to destroy foreign cells. No one has yet figured out how to do that well enough to match what happens during gestation, though. This is what "tissue-educated" means: that the neural stem cells were taken from fetal nervous systems. Aileen says that somewhere between sixteen and twenty weeks of gestation is the window during which they're most effective.

So that's the *nature* factor. What about *niche*? Specifically, is there a "best time" to do cell transplantation – a time post-injury when the environment at the location of the injury is most likely to let transplanted cells do some good? She's telling us the story of how there came to be a sort of conventional wisdom on that subject, and this is familiar to me. *The best time is the sub-acute stage, between a week and ten days post injury.* That's exactly what I heard, from the very earliest days after spinal cord injury came to live in our house.

It turns out that this idea came from a Japanese lab in 2003. Hideuki Okano was a careful scientist, and he made a good argument for the sub-acute stage as transplant heaven. Right after injury there would be a lot of inflammation, and the transplanted cells would be killed off. After a couple

of weeks there would be a glial scar in place, and the cells wouldn't be able to overcome it. The sweet spot would be sometime in between them. And, Aileen says, "*Every other person who did spinal cord injury research for the next five or six years did that. That is the time that they picked for transplantation, based on this completely logical argument.*"

What she's building toward is the rationale for leaving this conventional wisdom behind, though it seemed to make so much sense. One argument against it is that confining experiments to sub-acute injuries leaves every chronic out in the cold, and that's a lot of people to simply write off without even trying. Even if policy-makers can shrug at the monumental suffering, the cost in medical care and lost wages runs in the hundreds of billions. Another argument is that during the sub-acute phase, a lot of people recover some function spontaneously – which means that it would be hard to prove that cell transplants given in that window were doing anything. How would you know it was the cells and not the body healing on its own? Chronic injuries have stabilized, so by definition changes in function are not likely to be spontaneous. For the FDA, she says, the preferred standard for a chronic injury is sixty days post.

It's funny, in a way. Every single person who's sitting here in a wheelchair is way past sixty days post. Whatever the criteria is that makes a person chronic, in this room it was surely met a long, long time ago. Anyway. The experiment she wants to tell us about involved varieties of mice and rats that have been bred to have no immune system. They like to use these animals because the transplanted cells survive – and because, as they say, it's "clinically relevant." *Clinically relevant* is scientist-speak for *it's-more-or-less-what-we're-going-to-have-to-do-to-people*. The people who would get these kinds of transplants will need to take drugs that suppress their immune systems long enough to let the cells survive and start working. Therefore, rodents with no immune systems are logical surrogates for the early testing.

They started with the specially bred, immune-system-free mice. First, deliver the injuries. Wait for the supposed sweet spot, nine days post. Next add some human neural stem cells in very, very small quantities. Use four injections, like four push-pins holding a child's drawing to a corkboard. Two above the injury site and two below, extremely tiny so as not to cause any more damage. Record all kinds of data about how the mice behave and function. Sacrifice them, make super-thin slides of their spinal cords, stain them, and use high-powered microscopes to look at what happened.

So after all that, what do you see? She shows us. On the slides she's pointing to, you can see the cells clearly. Those cells knew what to do in

that damaged spinal cord. Remember that these were human central nervous system stem cells, snatched up just at the time and place on the branching tree where they were waiting for further instructions. You'd expect that – given the opportunity – they'd go on to complete the branching, and that's exactly what happened.

Two out of three of those little buggers became oligodendrocytes, and they did what oligodendrocytes always do: sent out tendrils that wrapped themselves sweetly and tightly around any naked axons in the neighborhood. A quarter of them went the neuron route, sending axons out to find some friends and establish a communication network. The whole thing knocks me out even if it never turns into any kind of treatment for anybody. This woman figured out how to put human fetal brain cells into a broken mouse spinal cord, and then the mouse remembered how to walk.

Yes. The mice regained the ability to walk in a coordinated way. We have a little movie of some of them doing just that. A good next question would be, how did that work? Did the human cells do all the recovery heavy lifting, or did they somehow cue the surviving mouse cells to get back in the game? It's important to know exactly what went on if this is ever going to be a therapy for humans. Aileen's team had a genius plan to answer that question:

poison those human cells while leaving all the mouse cells alone.

She did this with a diptheria toxin that's a hundred thousand times more deadly to humans than to mice. First they practiced in a dish, making sure that it would really kill those neural cells dead while leaving all the mouse cells alone. Then they gave it to animals that had been spinal-cord-injured, transplanted with human cells, and seen to have reached a recovery plateau. Here's what she says: *And what we saw is that those animals lost their recovery of function.*

So that answered that question. The human cells were what made the mice get better. Yesterday at one of the breakouts, Os Steward said that people who give grant money to scientists are always looking for something new, something that hasn't been tried before. He said that it's much harder to find the dollars to pay the cost of grinding out the series of experiments that you must do in order to fully explore all the features of a discovery. I'm thinking of that comment while Aileen describes what her group did once this mouse experiment was finished.

Before I go there, though, she'd want me to say that this experiment involved the close cooperation of dozens of people and more than ten years of effort. She didn't even start out to work on spinal cord injury; she was focused on Alzheimer's. And then the people at the Reeve Foundation talked her into

joining our fight and gave her some money to get started. Her husband and partner, Brian Cummings, is also a neuroscientist — in fact, he's the one who worked out the original transplant methods that made the whole thing possible. It's easy to forget, listening to someone give a talk like this, that the speaker is really telling the story of a deeply collaborative effort. She's telling us a sort of family history — it's about her husband, about the woman named Rebecca who manages her lab, about the raft of students who carry out the work, document the results, and eventually push the science forward. All those people are at the heart of the story she's just told, even if from our side of the screen it seems to be just a recitation of details about one more experiment.

So what did this team do once they'd finished that mouse experiment? They repeated it. Again, and again, and again. They did it with injuries at longer time frames, including chronics. They did it with different animals and different injury models. They did it with rats, because remember, rats are like humans in that when they get a spinal cord injury, it comes with a cavity that must be somehow bridged over. They showed that these neural stem cells work in animals. They showed that the old conventional wisdom about transplants failing in chronics was wrong. They showed that there is no sweet spot of a sub-acute phase for transplants. They showed that the cells they slipped into those injury sites got right down to business, branching off and becoming new nervous system cell types. Wouldn't the next logical thing be to try this in humans?

Why, yes. And that's what the next talk is about.

chapter**Fifteen**

Stephen Huhn: Human Trials for Chronics

"While much more clinical research needs to be done to demonstrate efficacy, the types of changes we are observing are unexpected and very encouraging given that these are patients in the chronic stage of complete spinal injury."

Armin Curt, M.D., Professor and Chairman of the
Spinal Cord Injury Center at Balgrist University Hospital,
University of Zurich

If you follow what happens in labs in the spinal cord injury world, you get sort of used to hearing people say that *much more research needs to be done*. It's always the same: we need to understand this a lot better before we'll be ready to take it to human trials. We just don't know enough yet. There are still a lot of questions to be answered.

What we're about to hear is not that.

We're about to hear about how a company based in Northern California took Aileen Anderson's work with animal models and translated it into safety trials with human beings. Those human neural stem cells have already been put into the bodies of three patients. The man who came here to deliver the latest news about those patients is Stephen Huhn. He used to be a pediatric neurosurgeon and is now a Vice President at the company that made this happen. It's worth taking a minute to build a little timeline.

- **1998:** A woman named Nobuko Ichida joins a young biotech company called StemCells, Inc.

- **1999:** Nobuko isolates human neural stem cells, the first time this has ever been done anywhere. Everything that follows is the result of her achievement.

- **2002:** Aileen Anderson goes to work at the Reeve Irvine Research Center. That same year, StemCells, Inc. approaches her and asks her to test their human neural stem cells in her lab.

- **2005:** Aileen publishes the results of her work with the human neural stem cells; the mice and rats are walking.

- **2007:** Stephen Huhn joins StemCells, Inc. as a Vice-President. One of his tasks will be to direct the process of getting Aileen's

animal experiments to human trials.

- **2011:** Three people with spinal cord injuries are given human neural stem cells produced at StemCells, Inc. in a procedure that closely mimics Aileen's experiments with rodents.

So, that's a super-quick summary of what it took to get to this moment, and it is by far the fastest basic-science-to-concept-to-human-trials path I've ever heard of. Ten years, in this world, is almost never enough time to get from basic idea to human testing. Back to Stephen. Based on my limited (but intense!) experience with neurosurgeons, he's easy to see as the sort of man who could operate on the brains of children. People who can muster the confidence and courage to do such things seem to share a particular combination of deep patience and awareness. They make me think of lions – deceptively sleepy and calm.

The very first thing he says is that we're only here today because all of the groups that had a part to play, played that part. These clinical trials would not have happened without basic scientists, universities, translational experts, private companies, ethical oversight boards, hospitals, surgeons, regulatory bodies, willing patients, and advocacy groups. All the pieces were needed. All the pieces were there. He goes on to show us a list that captures the history of transplantation as a part of human medicine.

It started in the 1950s with kidney transplants and expanded every few years to include more and more organs and body parts. He's also got a list of central nervous system conditions: MS, Alzheimers, stroke, Parkinson's, spinal cord injury, and many others. There won't ever be a whole brain or a whole spinal cord transplant to repair those things, but there could be transplants of the cells that are their component parts, and that's what he's working on. Spinal cord injury, he says, is especially intriguing to neurosurgeons because it seems so doable, so within reach. Just a couple of inches to repair for significant recovery.

StemCells, Inc., he says, is a publicly traded company. They've worked very hard to collaborate with regulatory and ethics groups all along the way, recognizing that stem cells mean controversy. It sounds like the approach is to always be in front of potential roadblocks, which you have to admire. We need people who have the foresight and political skill to overcome hurdles and get a treatment working. In the end, overcoming objections to cell transplants is going to depend on success, just as overcoming objections to "test tube babies" depended on the birth of that first healthy little girl. (That's my take, not something he said.)

Louise Brown's parents had been trying unsuccessfully to have a child for nine years when they agreed to become the first couple to conceive a baby through IVF in 1978. Since then, more than five million children have come into the world through that procedure.

So, how did StemCells, Inc. blaze this path? They've done other trials with their neural stem cells, but always on fatal illnesses that attack children. The FDA allows this kind of human trial much more readily than it would allow a trial on patients who are "merely" paralyzed. And what they learned from transplanting cells into the brains of those babies is that the cells survive and migrate deep into the brain. They learned what sort of dosage could be tolerated, and they saw signs that the cells became oligodendrocytes and built myelin. The cells didn't save the kids' lives, and there will be another round of trials – but this was information that helped them make a case for moving on to work with spinal cord injury.

The trials for spinal cord injury aren't happening in the USA; they're being done at a university hospital in Zurich, Switzerland. Stephen doesn't say exactly why, just that one of the reasons they chose to do it there was that the University of Zurich has an extremely well-defined way of objectively measuring outcomes. I'm listening to this and thinking of all the times I've read about stem cell transplants in other countries (Mexico, India, Brazil, China, for example). Those transplants are not what Stephen is talking about. The source of the cells in those transplants is often unclear. The science behind the methods is murky. The "proof" that the transplants work is usually just what the patients report feeling. It's reassuring to hear in such detail exactly what Stephen's company is doing, why they're doing it that way, and how they plan to measure success.

Here's where we are. StemCells, Inc. asked for volunteers who were between three months and one year post-injury and who had been diagnosed as ASIA A, with no sensation or movement whatsoever below the injury level. The patients had to have thoracic injuries, between T2 and T11. The people selected to be in this first group included three men, aged between twenty-three and fifty-three. Each of them was given drugs to suppress their immune systems, and each of them had surgery. The surgery involved exposing their spinal cords and using a very tiny needle to deliver 20 million human neural stem cells into four locations just outside the injury site, exactly as Aileen's team had done with the mice and rats.

All the surgeries happened in the last few months of 2011, which means that Stephen's talk is timed at about a year post-transplant for these first patients. These guys were not expected to get any return, but *two of them did get measurable return of sensation*. These were guys with complete, chronic injuries. None of them had increased pain or any other kind of negative results. Stephen goes methodically through the data for each patient, carefully explaining how

> ASIA isn't about geography; it stands for American Spinal Injury Association. This group defines the four classes of injury: A is the most severe, and D is the least. ASIA A patients have neither movement or sensation at the sacral level. These injuries are often called "complete." ASIA D patients have at least half the muscles below the injury level working at a score of at least 3 on a scale of 1 - 5. To get a 3, the patient must be able to use the muscle with full range of motion against gravity. He doesn't have to lift a weight – just raise his own arm or leg.

each of the University of Zurich sensory tests work and showing us why they got permission to keep going with these trials.

That's the good news; they're going forward. Twelve months after the surgeries, they definitively established safety, which was the purpose of this first effort. The selected dose of human neural stem cells did no harm and actually had a measurable effect, even in these worst-possible-case complete injuries. He shows us a graph that illustrates how unexpected any change at all is when the diagnosis is ASIA A. For almost all complete patients, nothing happens at all – but for two of the three complete patients in the trial, the transplants beat those odds. Their function changed. As I write this in March 2013, four months after the conference, the StemCells, Inc. website is actively recruiting volunteers for the next phase of the trial.

What will that look like? The next cohorts won't have complete injuries. They'll be classified ASIA B and ASIA C. They'll be from Europe, the USA, and Canada. They'll all come into the surgery with some signal getting through the injury site, which means – in theory – that the neural stem cells will have more to work with. They'll encounter a situation where there are spared axons looking for myelination. They'll find healthy neurons trying to form synapses.

I'm rooting for them – for the transplanted cells, for the patients who volunteer to trust this science, for Aileen's team of researchers, for the investors at StemCells, Inc., for Jonathan and everybody at CIRM, and for our community as a whole. It's time for a win.

chapter Sixteen
Considering Combinations: Still More Questions Than Answers

They elevated themselves by their own self-education on these things. And then it became very clear that you weren't going to mess with these people, because they knew exactly what you were talking about -- and they knew exactly what they were talking about.

Dr. Anthony Fauci, Director of AIDS Research

How to Survive a Plague

Before we can make sense of this Q and A session, it might help to have a little more background about the kind of trials that StemCells, Inc. are doing in Zurich. There's a standard way that new drugs get brought into general public use, and it starts with what they call "pre-clinical" work. Pre-clinical just means not-ready-for-humans; it's animal studies. Aileen's lab did the pre-clinical work that led to Zurich. The next stage is called Phase I, which is when, typically, a very few volunteers get a very small dose of the drug.

There's no expectation in a Phase I trial that the small dose will do any good; the idea is to make sure it doesn't do any damage.

Phase I trials are a bit like taking a very tiny taste of a new food, just to make sure it's not going to make you sick. You're not actually expecting that small bite to fill you up. Phase II trials happen next, and that's when the dose gets increased, often gradually, to keep checking for safety and to start looking for benefits. The number of people tested is still kept very low. If all goes well in Phase II, you go on to Phase III, where more people will take the drug at doses expected to have benefit, and the outcome targets are very well-defined.

The Zurich trials are a hybrid of Phase I and II. The number of patients is tiny, a reasonable dose has been selected, and safety (no harm done) is the primary target. The doctors are, however, also measuring changes in sensory and motor function as they go, which is why they refer to these trials as Phase I/II. The other thing to know about these trials is that there's no control group. In most tests of new medicines, researchers will have Group A getting the drug and Group B getting a sugar pill. They keep the study "double blind," which means that nobody – doctors, scientists, or patients – knows who is in which group.

The Zurich trial isn't like that. Everybody knows who got the surgery, right? It would be a pretty hard thing to fake. The patients' recovery isn't going to get measured against a set of controls, but it is going to get measured against the whole population of people with similar injuries and histories. That's what Stephen meant at the end of his talk when the said that two out their first three patients were outliers in terms of recovery. It's well-known how much recovery to expect from a chronic ASIA A injury: most of the time, none at all.

The first question for this panel comes from Donna Sullivan, who wants to know if the Phase III part of the trials is going to involve rehab. Stephen says yes, for about six to eight weeks, and not just in Phase III. They're going to have all their ASIA B and C patients do a uniform course of physical therapy in the Phase I/II trials, too. The patients will need some just to recover from the surgery, and of course physical therapy is going to be part of any mobility gains after a long time of being paralyzed.

Jerry Silver wants to know more about the placement of the cells and the particular kinds of sensory recovery that two of the patients got; did it matter exactly where those injections were? Aileen says that in the mice and rats – which of course they later dissected and examined under massive magnification – those cells migrated throughout the spinal cords. You couldn't even tell where the original injections had been made. Obviously they can't be sure the same thing is happening in people, but there's no reason to assume it isn't. Stephen says that it's one of the things they're looking for in patterns of recovery, to see if the location of the injection seems to make a difference in the outcomes. I'm knocked out again by how many variables there are to consider: size of dose, method of delivery, place of delivery, quality of rehab, type of injury, and on and on.

Bob Yant asks for a microphone. He wants to know if Aileen's lab changed their strategies based on time since injury. Did they do the same experiment no matter how old the injuries were? She says they did a lot of models, all exactly the same: same dose, same locations, same followup. She also tells us something she didn't mention earlier – that there was only one group that got no benefit at all from those neural stem cells, and that was the acute injury group. The neural stem cells didn't help when they were injected immediately after injury.

She goes on to say that they're also moving into cervical models now, which I'm glad to hear. It sounds like her lab and StemCells, Inc. have formed a solid back-and-forth sort of collaboration, where she figures out what's likeliest to work in humans, and they take that knowledge and put it work in trials. This seems like a painfully obvious plan, but it's not something that happens very

often in spinal cord injury – maybe twice in all the years I've been watching.

And then she and Stephen riff for a few minutes about how very tough it's going to be to get to trials of combination therapies. It makes my head hurt to even imagine. Aileen: *Everything has got to be validated separately for safety, and this is sort of a sticky wicket in terms of moving things forward . . . to do that – to combine with chondroitinase or those other interventions that people are doing – is going to require an awful lot of preclinical safety data. And it's hard . . .* Stephen: *Yeah. You don't have to look at the field for very long to realize that there's going to be power in a combinational approach. The issue is translating that into a pragmatic trial . . . I think it's likely that the cell will address – if we get there – some aspect of the disease or the injury, but not all. So that's where the combinational questions come in. And it's fascinating to think about, and we can all on paper think about studies that we would do. The problem is in the execution.*

Sigh. They're both saying the same thing. We're going to have so many variables, each of which will need to be tested separately and then again in combinations. What's the most efficient way to do that? At this point no one knows.

Someone from the audience wants to know if they've thought about using InVivo's scaffolding product to deliver the cells. Stephen says for sure they're tracking that technology, and that it gets back to the question of combinations. Using scaffolds would be about the possibility of being able to do more than one thing at a time. Aileen agrees with that – and she says again that this issue of how to combine all the possible therapies and delivery systems is going to be challenging. Listening to them, I'm reminded of the AIDS activists who struggled first to get attention, then to get whatever kind of drugs were available, and finally helped design combination therapies. The combinations worked, but it took *years* to get them – and their friends were dying by the thousands while they tried to figure it out.

Another audience member wants to know why they didn't do primate testing to learn more about what might happen when they got to humans. Couldn't some guesswork have been eliminated in terms of dosage and injection sites? Aileen is emphatic: *No*. The problem has to do with immune systems. She got into this neural stem cell thing in the first place because she had made genetically altered rodents without immune systems. The benefit of those rodents was that you could put human cells in them, and the cells would stay alive. There are no immune-system-free primates, which means that if you wanted to put human cells into, say, monkeys, you'd have to somehow keep their immune systems from attacking the cells. And that – in her opinion – would be a lot more trouble than it's worth.

Stephen chimes in to say that the question behind this question is about what's enough pre-clinical evidence to make it worth the time and expense of going to trial with humans. A lot of things that work in animals fail in people, no matter how convincing the data is beforehand. Every animal model is going to be different from humans in some way, so in the end every scientist who wants to do translational work has to make a judgment call about which animal models will provide evidence that's good enough.

Someone from the audience wants to know if Stephen's team was able to look at the MRIs of their first three patients to see if the sensory recovery was due to myelin regrowth? He's asking, basically, if they can prove that those transplanted neural stem cells turned into functioning oligodendrocytes. Stephen says, unfortunately, they can't look at that, even with very sensitive MRI technology. The patients all have hardware around their injury sites, because the broken bones in the spine have to be fused with titanium rods and plates. The hardware interferes with the MRI.

Okay, weird personal story that I can't resist telling here. My husband had a small titanium plate with four screws attached to his vertebrae after his injury. When he was four years post, the screws somehow worked themselves loose from his bones. The plate – with all four screws still attached – migrated through his body toward his throat. A pouch of flesh formed around that plate, and naturally it interfered with his ability to swallow. Not knowing why he couldn't choke down his pills, he went to an ear-nose-throat guy, who sent a scope down his throat to have a look.

Big freakout! They sent him straight to the trauma center where he had been an inpatient and arranged to do surgery to remove those foreign objects, which obviously didn't belong where they were. There was huge concern about infection, and nobody could understand why he hadn't lost function. Didn't titanium floating around in the cord damage more tissue? Apparently not, because his only symptom was trouble swallowing.

There was a whole team of surgeons assembled to do the extraction, and it was supposed to take a couple of hours. There would be a few days in ICU and a long recovery period. Thirty minutes after they started, the lead surgeon came to find me in the waiting room. All done. Looks good. He can go home tomorrow.

This is where you need to know that my husband has a big mouth. Literally, a wide smile and a long jaw. The surgeon had decided — before cutting his throat open – to see if he might be able to get a visual on the plate first. So they opened my anesthetized husband's big mouth as wide as it would go . . . and

realized they might be able to just use a good pair of tongs to pull that plate out. And that's what happened. I'm thinking of this when Stephen says they can't use MRIs to see if the cells made myelin. *Hey, Dr. Huhn! I know this guy who's got a chronic injury and no hardware at all.*

Ahem.

The session ends with Donna's reminder that the Zurich trial is open to people from Europe, Canada, and the USA. They're actively recruiting volunteers with thoracic injuries who are between four months and one year post, and whose ASIA scores are either B or C. Good luck, y'all. Go for it.

Bridge Builders

chapterSeventeen

Jerry Silver: Bypass the Whole Damn Thing

Happy is he who gets to know the reasons for things.
Virgil (70 - 19 BCE)

Now comes a very happy-looking Jerry Silver back to the podium for his second presentation of the conference. He starts with a quick story about what goes on inside the cord at the moment of injury, something I have to admit I really hate to think about. It's like watching a crash in slow motion, hearing this description and knowing the outcome in advance. Here's the deal: our blood is not supposed to come into contact with our brains and spinal cords, and that's what happens during the injury.

We all have a built-in filter called a *blood brain barrier* that protects our central nervous systems from the harmful bacteria or viruses that routinely

travel through the rest of our bodies. Scientists figured this out more than a century ago while they were trying to create dyes that would make it possible to see specific biological structures under microscopes. They put blue dye into animals' bloodstreams and saw that everything *except* the spinal cord and brain turned blue. Later they put the dye into the brain, and then *only* the spinal cord and brain changed color.

When blood enters the brain or the spinal cord after trauma or stroke, therefore, what the brain sees is something completely unfamiliar – a foreign object. The immune system has a standard reaction to foreign objects: it tries to get rid of them. And so a horde of little munching cells called *macrophages* (the word is Greek for *big eater*) stampedes to the injury site and starts devouring everything in sight. They eat up whatever debris and injured axons are there as fast as they can, and even inflict damage on the remaining healthy axons. In the meantime, reactive astrocytes are hustling to block off the injury site from the rest of the cord. They're making the scar, and it's good that they do that, right? If they didn't, the macrophages would just keep eating up more and more axons.

So. We have a scar, and we know that inside it, right in the core of the lesion, are a bunch of cells called oligodendrocyte precursors. If they could, those cells would turn into oligodendrocytes and make myelin. They don't do that, though. They just sit in the injury site, teasing those struggling-to-grow axon stubs into a state of permanent entrapment by encouraging them to make aberrant synapses. This is the situation Jerry described for us yesterday: the beach chair scenario.

The precursor cells are surrounded by the second wall of defense, that tough glial scar. The task is to get all those axon stubs off of the glial precursor cells so they can grow through that cavity and beyond the outer scar walls. Not an easy task. What if, Jerry says now, maybe the best plan is to *build a bridge and bypass the whole damn thing?*

It makes sense to build a bridge and give the axons something to grow on, like a living trellis. You'd still have to deal with the fact that they're not in a growth phase of their lives and not happy when they're anywhere near the scar, but that's what the combination therapy idea is all about. Ramon y Cajal actually figured out how to use *peripheral nerves* to make a bridge inside a damaged spinal cord a long, long time ago, except nobody believed he'd actually done it. Then in 1981 a couple of other scientists repeated his work, and they had something Cajal didn't: axon-tracing techniques. The tracing techniques let everybody see that Cajal had been right; he could get axons to grow into a bridge.

The problem was that no matter how many axons grew happily and

Fingers, wriggle! What's supposed to happen when you have that thought? The neuron cell bodies in your brain push the order down through your corticospinal tract. Then another set of neuron cell bodies sends the message out of your cord and into the muscles of your fingers. The axons that take a message from cord to finger muscles are in bundles called *peripheral nerves.*

for long distances *into* that bridge, hardly any would come *out*. And that, Jerry's lab finally figured out after twenty-five years, was because of those molecules called proteoglycans. So now the combination strategy begins to take shape. A bridge of some kind that bypasses the injury site, plus some chABC to help break down the scar, plus some growth factors for nourishment, plus, plus, plus – it's like an action game, where new kinds of enemies keep rising up out of the ground. You have to beat at least some of every kind in order to win. You have to make sure your tactics don't cause new kinds of enemies to be created. And you have to make sure you don't damage any of your own armies in the process.

Jerry was looking to do something dramatic. He'd already shown that by combining chABC and a bridge, he could restore breathing in a rat with a hemisection injury. He'd gotten a single muscle to work again. He wanted to tackle a harder puzzle, and that's why he decided to do a complete transection and then see if he could grow axons across it and restore – this time – two muscles. Which muscles? How would this experiment be designed so that there would be absolute certainty that function had been restored by having axons grow past the injury site and connect to muscles?

* * *

Back for just a minute to the balls and strings. We said earlier that the neuron cell bodies in the brain are grouped into little collections, depending on what sort of function they control. The ones in the part called the motor cortex give rise to tracts of axons that eventually send the signals that let you walk. But there are other tracts, other bundles of axons originating in other parts of your brain, and they're the subject of this talk. We're going to hear about a section of the brain called the brainstem and the *pontine micturition* center, where certain tracts of axons originate that are important for bladder function. (*Micturition* is Latin for urination. Of course.) These axons, as it happens, follow a molecular growth path that's more primitive and different from corticospinal tract axons. This matters because they don't need PTEN knocked out in order to grow. That's what we're going to hear about. Also, we're going to hear about how exactly a bladder operates. We're going to hear about pee, specifically about rats and their oh-so-interesting peeing behavior.

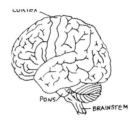

You can think of a bladder as a little empty sack with walls made of muscle. It has a tube going into it and a drain at the other end. The tube allows pee to drip in all day and night as it slowly gets produced. At some point the little sack gets full and must be emptied. This takes a coordinated effort; the bladder walls have to squeeze just as the drain gets opened. If the drain opens and there's no squeezing, pee will leak out in annoying little drips. Opening the

drain requires relaxing a muscle called a *sphincter*. Squeezing the bladder requires contracting a muscle called the *detrusor*. You need both squeezing and relaxing, both detrusor and sphincter, and you need them synced.

Jerry tells an intriguing (and deeply frustrating) story about why he decided to go for these two muscles. It starts with a couple of scientists who published a paper in 1996 that said they'd succeeded in getting rats with complete cord transection to walk by creating a peripheral nerve bridge. It would have been a huge deal if that had been true, but it wasn't exactly. They exaggerated, or maybe just convinced themselves that they'd really done it. Either way, nobody could reproduce their results, and so the conventional wisdom became *bridges don't work*. The NIH stopped funding scientists who wanted to try them. This one false claim turned the field away from what might have been a promising therapy.

One of the people working in the lab that produced those experiments ended up in Jerry's lab in Cleveland, and Jerry tells us that he asked this man, whose name is Yu-Shang Lee, if any of the rats had really walked.

"So I asked Yu-Shang, 'Do these animals walk?'

And he admitted to me, 'No. But they sure can pee.'

I said, 'Really? That's very interesting and very important. Let's study that.' And so we did."

The reason it was very interesting and very important, Jerry knew, was because it was evidence that they had in fact succeeded in growing axons and restoring some function – just not the function they were claiming. And so he set out to do the experiment with restoring bladder function. He thought he could make the results even better than before, because he had something the first team hadn't: chABC.

He describes the experiment. Do a complete transection at T8-9, get some *intercostal nerves* from the same animal and soak them in chABC. Stitch them across the injury site, adding more chABC on either end of the bridge. Put fibrin into the cavity after juicing it up with a growth factor called FGF. He did that. And what he found was that when he combined all three pieces – the chABC, the FGF, and the nerve bridge – a lot of axons could grow all the way from the brain to the bottom of the cord. But – and this is why we needed to know about the different sections of the brain – only some axons will respond to this treatment. Which ones?

Intercostal nerves are bundles of axons that come out of your chest-area spinal cord and wrap around your ribs.

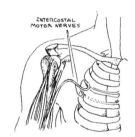

The primitive ones that originate in the brainstem and go on to connect to the two muscles that control peeing. Only those. They can regenerate and get across that bridge. They don't have a more modern genetic recipe telling

them to never grow again; if they did, we'd have to find the gene and knock it out before getting all that growth. The axon tract that begins in the cortex and controls walking doesn't work that way.

Jerry spends some time showing that the triple combo treatment led to rats peeing in something close to their normal patterns. Did you know that when rats pee they don't usually do it all in a rush, like us, but rather in little pulses? And did you know that they normally pee every half hour or so? Me neither. It's a fact, though, and the reason I care is that Jerry's T8-9 transected rats got that behavior back after getting the triple combo. Their squeeze/relax muscles worked again in a rhythm that the untreated rats could only dream of (if rats could dream).

How would you go about making sure that this recovery was really the result of the treatment itself? One way would be to cut through the bridges and see if the rats went back to not being able to pee normally. Jerry's team did that, and the rats lost those two muscles again. Good evidence that the bridge was a necessary part of the recovery, right? Another way involves interfering with a couple of *neurotransmitters* known to be important in bladder function. Neurotransmitters are part of the information chain; remember? They're chemicals that get released by neurons about to form a synapse with other neurons. (I think of them as something like smells, like the smell of baking bread in my kitchen, or the smell of garbage. They cause reactions in me just by filling up the air around me.)

The point is that these chemicals are a necessary part of the chain of information that goes from the brain to the squeeze/relax muscles in the bladder. If the whole chain is working naturally, those chemicals have got to be there. If you could block just these special chemical transmitters in a treated rat and see the restored function disappear, you'd have more evidence that the bridge had established a true channel from brain to bladder. When Jerry's team functionally blocked two neurotransmitters *serotonin* (by blocking its receptors) and *noradrenaline* (by blocking its synthesis) they saw just what they expected – the bladder recovery was wiped out.

So that was two pieces of evidence. The last one was the bladders themselves, examined after the animals were sacrificed. If the treatment really worked, you'd expect those little sacks to look like the ones in normal rats and not like the ones in injured rats. Injured rats have thick, tough bladders, and normal rats have very thin ones. The treated rats had thin ones, which is yet more evidence that the triple combo had done what's supposedly *not possible*: caused axons to grow past a complete transection, make functional connections on the other side, and restore muscles.

*   *   *

Does it work in chronic injuries? I will listen to lectures about the finer points of rat pee and bladder thickness all day long if it leads me to a positive answer to that question. I can't say it better than Jerry did:

That's acute injury. That's where we started. And we thought, all right, if it works in acute stages, maybe we can apply this to the chronic injury. That's taken us one year, and here is the data. This is all brand new. What we've done – and this is all a credit to Yu-Shang – is to contuse the spinal cord with an Infinite Horizon device as hard as we can. This is not a complete transection, it's a contusion lesion, because that's more like the typical human condition.

And then we decided to bridge the lesion exactly the same way we did before, with our triple combination: FGF + bridge + chABC. And we went from lesioning, a period of two months passes, then we graft . . . just like we did before. We lesioned, we grafted, and we waited, hoping that we'd see changes in bladder function or anything.

And there was nothing. As a matter of fact the animals got far worse. And it was pretty depressing. After six months of failure we had a long discussion, and we thought, okay, let's change our strategy and add this intermediate phase. We call it the wound preparation phase.

They didn't give up. They believed in their science, so they devised a new strategy and started over. The animals he's talking about right now are still being studied. Those rats were given a massive contusion injury – not a transection but the smashed strawberry injury that's so devastating. Jerry's team delivers that injury and then waits a couple of months so the animals are in a chronic phase. They go back in with what I think of as a sort of "softening up" goop, consisting of fibrin spiked with chABC and FGF, and they fill the injury cavity with it. They add a couple of shots of chABC above and below the injury, close the wound, and wait a week.

Now it's time for the bridge. Same exact process from here on out that they used on the acutes, and this time – so far at least – *it's working*. They're seeing thousands of axons entering and exiting the bridge in a chronic injury, contused cord model, and those rats are getting back the squeeze/relax muscles that let them pee more normally.

So, how to build on this?

The possibilities practically bounce out of him: add the peptides, add the PTEN blockers, add more growth factors, use that salmon fibrin, do different injury models, wait a whole year before grafting, anything we can think

of, we should try. He ends with a quote from Cajal, who has been proved right about so many things, but not this time: "*Once development is complete, the sources of growth and regeneration of axons and dendrites are irretrievably lost. In the adult brain the nerve paths are fixed and immutable; everything can die, nothing can be regenerated.*" (1928)

Not so, says Jerry. Cajal was wrong about that, and Jerry Silver delivers a line like he wants it to be quoted. "*Silver, Working2Walk 2012: After complete cord transection, even at chronic stages, long distance regeneration with clinically relevant functional recovery is possible.*"

It's about time.

chapter**Eighteen**

Justin Brown: Low-hanging Fruit

I'm motivated again. I've been slacking a little bit, but I'm motivated again to get out there and do a little more. Get myself out of this chair.

Advocate Karen Miner

Is there anything that can help any of us now? Right now? Not in a couple of years, after all the animal testing of combination therapies leads to more and more human trials, but this week, now, today? Dr. Justin Brown is a reconstructive neurosurgeon at the University of California San Diego Health System, and he's here to say that the answer to that question is *yes*. Specifically, there's a procedure that can restore hand function to people with chronic injuries at C5 and below. Seriously. If you have control of your biceps but can't move your fingers, the doctor standing in front of us now is one of the few people in the world who have the skill and knowledge to get your fingers moving.

How is this possible?

It has to do with peripheral nerves and their ability to heal. A little anatomy: we use the word *nerve* to refer to long, slender tubes – that is, we call these things *peripheral nerves* when they're in your arms, legs, hands and feet, but when they're in your spinal cord we call them *tract*s. Inside those tubes/nerves/tracts are *fascicles*, which are just discrete bundles of axons. (The scientist name for the word, *bundles* is *fascicles*, which is the Latin word for bundle. Go figure.) The fascicles are part of the grand communication network that goes brain-to-spinal-cord-to-everything-else-and-back-again. Peripheral nerves are programmed to repair themselves when they're damaged. Justin describes the process this way:

> Schwann cells are to the peripheral nervous system what oligodendrocytes are to the central nervous system: they make the insulation that axons need.

As soon as a peripheral nerve is cut, those Schwann cells start multiplying and they line up and they start throwing out breadcrumbs. They want those axons. The axons begin to sniff those things out, and they start heading down towards the target. This environment was set up to re-grow, so that when you have an injury in a limb, you will recover after time.

The issue in the peripheral nervous system is the train tracks. The train is the axon, the tracks are the nerve itself, the conduit for those axons to grow down. And as long as I can put a reasonable train track back together in a reasonable time, the axons will make their way to a target.

"The train is the axon, the tracks are the nerve itself." What he's talking about is the tubular structure that holds the bundles of axons. Unlike spinal cord axons, peripheral axons always grow back after they're damaged, as long as they have a structure to grow into. They never have genetic instructions telling them to refuse to grow. Justin shows us some pictures of people who have no spinal cord injury but whose arms are paralyzed and useless because their peripheral nerve axons went off the rails, so to speak.

He learned to do nerve transfers working with these folks, usually people who had terrible shoulder injuries after motorcycle crashes. He would do a surgery that involved taking a bit of healthy nerve from somewhere else in the patient's body and transferring it oh-so-carefully into place where the original train tracks had been destroyed. The pieces he's stitching into place are one millimeter in diameter, the size of the wire in a paper clip. This seems like it would be self-defeating, right? Aren't you going to take away function somewhere else if you snip out a section of nerve?

Not really. Our bodies come with a certain number of spare parts. In particular, we have lots of nerves duplicating each others' work and doing the same basic job, so it's a matter of figuring out which ones can be borrowed and re-purposed somewhere else. The most effective nerve transfer operations happen, says Justin, when the nerve you're borrowing for a new use is physically close to the one that's getting repaired, and when the function you're trying to restore is simple.

One downside of this kind of treatment is that it takes a very long time – six months to a year – for those axons to grow through a small re-directed chunk of nerve. Another is that it's very labor-intensive to retrain a nerve; you have to want it, and you have to work at it. For people looking at an indefinitely long future with no hand function, though, what's a year? Justin's first spinal-cord-injured patient was twenty-eight years old. He'd been a C5 quadriplegic since he was fifteen, and he wanted his hands back.

The surgery makes sense. The reason this young man's hands didn't work is that messages from his brain were not able to get through the injury site, through the spinal nerves, and out into the peripheral nerves that are there to signal his finger muscles to bend and straighten. The peripheral nerves were intact. The finger muscles were intact. All that was needed was a signal from somewhere. There was a perfectly good signal getting from the brain to the nerves in the biceps, and those nerves became the donors.

Specifically, what Justin did was open up the nerve that feeds information to the bicep muscles and split out a few fascicles from the many bundles inside that tube. He attached those fascicles to a segment of what's called the median nerve. The median is right next to the bicep nerve; it delivers information to the forearm and hand. From the perspective of the median, this was a new injury, and the axons inside that nerve did what peripheral axons do: they followed the Schwann cells' breadcrumb trail and created a new train through this newly laid track.

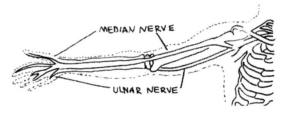

What happens when axons that start out headed for bicep muscles end up in hand muscles? That's the most interesting part of this talk. Rehab after this kind of surgery involves a lot of waiting while those axons slowly, slowly grow to the waiting muscles and attach themselves. The growth rate is only one millimeter a day, so this takes a very long time. Once it's done, though, the rehab can start. The patient thinks, *bend, fingers!* And nothing happens. The patient thinks, *bend, elbow!* And the

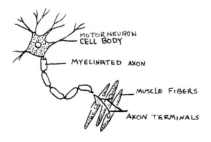

elbow does bend . . . but the fingers flex, too. Some of the axons that used to bend the elbow have been diverted so that they're landing on finger muscles.

Justin says that if the patients work at it, the brain very quickly reorganizes itself into a new map that takes advantage of the new connections; no conscious effort is required. His patients can flex and straighten individual fingers – they're not likely to play the piano, but they can use their own hands far better than they could before the surgery. He shows video of some actual surgery, which I will admit to looking away from almost every time. (It's in the video references in the resources section, so you can watch it if you want.) The part I did watch was the procedure for making sure they were going to attach to the exact nerves they wanted to bring back. With the patient anesthetized, his arm and median nerve opened up, the doctor would stimulate each fascicle and watch to see what the hand did in response, like a mechanic with a voltmeter. Touch *here*, and the sleeping patient's hand gripped the nurse's hand as if he were trying to tell her something. Bingo.

* * *

Does this kind of surgery work for people with tone and spasticity? Well, sort of. Justin says that when you open the arm and look at the muscle groups of someone with a lot of tone, what you see is a mixed bag. Some of the muscles are working normally, some are not working at all, and some are basically always "on." If you try to attach nerves so that the muscles not moving at all will start getting messages from the brain, it might not work. The reason is that often those muscles will be in an agonist/antagonist pair with the ones that are always on.

An example would be the pair of muscle groups that let you bend and extend your thumb; when one is working, the other has to be relaxed. The way they deal with this in Justin's practice is to do what they call *selective peripheral neurotomy*, which is a fancy way to say that they trim back some of those fascicles that are getting constant input from the brain. They can do it very, very precisely, so that there is still some message getting through but not the constant onslaught that results in tone.

We get to see a couple of video-taped examples of this. One shows a man who used a walker to get around for fifteen years and then had Justin detach some of the nerves that were causing his toes to point permanently inward. Afterwards he still couldn't tap dance, but he didn't need to hang onto the walker and drag his feet around, either. He could just walk.

Justin's making a pretty good case here. His attitude is that if people can be helped right now to have more independence and a better quality of life, that's a reasonable choice. Peripheral nerve transfers aren't going to interfere with anything that later becomes available to repair the cord itself, so that's

not an issue. This approach represents a sort of low-hanging fruit for people willing to deal with the surgery and the recovery and the rehab. It's certainly not a cure, but it's hard to argue that the twenty-eight year old man who had spent almost half his life with no hand function should have waited a little longer for something better.

chapter**Nineteen**

We Just Gotta Get Some Axons to Go

We live in a time when the words impossible and unsolvable are no longer part of the scientific community's vocabulary.

Christopher Reeve

It's now almost lunchtime on the second day of the conference. We've heard about genes, drug delivery strategies, advocacy, public funding, neural stem cell transplants, rats peeing, axons regenerating, being paralyzed in Nepal, beach chairs, moats full of sharks, and tactics to break down glial scar. We've been hanging out together, sharing meals and beers and trying to take in as much information as we can possibly absorb. For the record, I need to say that it's getting harder by the minute to keep it all straight. Jerry and Justin are up front, looking curious and eager. I'm thinking about how different this event is for them than for us.

They have context. All the presentations here fit into a bigger picture that they've been looking at for a long time — a sort of brightly colored, meticulously detailed painting. It's like there's a giant mosaic, and each of them owns a little piece of it. They spend their days working on their own section with their various instruments, and every little while they step back to see how the whole thing is coming together. For them, that's what a scientific conference is — a chance to look for new relationships between what they're doing and what others are doing.

That's what's about to happen now. Jerry starts things off by pointing out that Justin ought to be using chABC in his nerve transplant surgeries. The same proteoglycans that stifle axon growth in the spinal cord are there in peripheral nerves, so it only makes sense. Also, he knows of another researcher who has already shown that it works. Justin nods agreeably: "*Okay. Sounds good to me.*" Jerry's main response to what Justin is doing, though, has to do with how readily the brain re-maps itself to take advantage of new communication possibilities.

Right after nerve re-routing surgery, Justin's patients have to work hard to move the muscles that just got new nerve signals. If the newly connected nerve used to fire a bicep but is now attached to finger flexors, then thinking about curling a bicep will result in the fingers bending . . . until it doesn't. *That's the remarkable thing, that the wiring in the brain responds and re-configures itself to match the new wiring in the body.* Jerry keeps saying that this is just incredible, and so encouraging. To him it means that our nervous systems are ready to grab any lifeline we can throw them. It's not going to have to be a grand, red-carpeted staircase; a sturdy rope and a rung to climb on will do.

Justin is nodding again, pointing out that it's much harder for his patients with brain injuries to recover good function after a nerve re-routing surgery. "*It really is dependent upon somebody who's able to learn, able to develop some sophistication with what they do have before you can go to the next level. If you don't have good . . . control of the muscles that your brain does control, then rewiring them to something else is not going to work as well.*" Listening to him, I'm reminded of an adult student I once had. She could learn tricky concepts in calculus, but she couldn't hold onto them overnight. Every day – with tears of frustration – she'd have to start over.

There's a question from the audience about trying other nerve-bridging strategies to make the legs work, and after some confusing back and forth about why the suggested tactic is near anatomically impossible, Justin mentions that he knows of a surgeon in Italy named Giorgi Brunelli who has identified at least one way to do this. Brunelli spent decades in animal models until getting permission to ask for human volunteers. One of his patients was a woman with a chronic thoracic transection. He took chunks of nerve out of one of her legs and used them to create a new set of connections between her spinal cord and the three muscles needed to stand up and walk: the *gluteus maximus*, the *gluteus minimus*, and the *quadriceps*.

The surgery took twelve hours. One end of these transplanted nerves went directly into the corticospinal tracts in her cord, and the others were sent to the three target muscles. After a year, she started to get recovery in those muscles. Two years after that she was able to walk about thirty yards with a walker. What surprised everyone – and Justin is saying this now – is that nobody believed such a thing could possibly happen. The axons in the brain and spinal cord form synapses that involve one set of neurotransmitters, and the ones in the rest of the body involve another set.

A graft of axons that used to be part of the woman's leg, therefore, shouldn't have been able to make a meaningful connection to the axons inside the cord, and vice versa. The synapses weren't supposed to function with the wrong neurotransmitters, but they did. We know this because those re-routed nerves were able to fire the woman's muscles. It's new and important information, mostly because it shows that the nervous system still has secrets;

we don't know everything we think we know.

As a way to recover walking, it leaves a lot to be desired. How many people would sign up for a long and dangerous surgery, three years of rehab, and a limited result? Justin says that the limited result has to do with the weak power of the connection. The woman's legs are getting a signal, but it bypasses the usual route down through the spinal cord. When a signal goes from the brain directly to the muscle, it's going to be lacking a major "amplification" that comes from triggering the central pattern generator. The central pattern generator is a collection of neural networks inside the cord that takes certain signals and runs with them, so to speak. It's as if it hears a single note from the brain and turns that note into a whole symphony of music that the rest of your body is tuned for. Brunellis's re-wiring project skips that whole thing, and so his patient was only working with the single notes. She *really* had to want it.

There's another question for Justin about the possibility of doing a transfer from a firing intercostal nerve (one that wraps around the upper ribs) to one in the hip or the lower trunk, but Justin says that he's already tried that. It doesn't work very well, he explains, because in that case you'd need to borrow too many nerves to get function worth having. There would be significant loss of trunk strength and just not enough gain in the lower body to justify the pain.

There are a lot of muscles in the legs (glutes, knees, hamstrings, quads, etc.), and you couldn't get them all back just with intercostals. *"We could probably get it to move. But that's still going to be a tough one to do without making the cord better."*

Someone asks Jerry if he knows about Susan Harkema's trials. Susan Harkema's lab in Louisville, Kentucky has been doing experiments that combine electrical stimulation and aggressive locomotor-based physical therapy. Lots of people with spinal cord injuries use e-stim, which involves attaching little electrode patches to certain spots on their legs that make it possible to push the pedals of a stationary bicycle. It's a real workout in the sense that your own muscles are firing; they're just getting the nudge to do so from outside the body instead of through the cord.

What Harkema does is different. She puts that electrical stimulation into the cord itself, specifically into the *epidural space*, which surrounds the bones in your vertebrae. It turns out that when you put a tiny electrical charge into that space, it's "felt" by the cord, which is a couple of layers away. She's not stimulating the patients' muscles; she's helping the cord reorganize its own wiring in response to the sensation of stepping. Harkema's first patient for this procedure is named Rob Summers.

Rob was the victim of a hit and run driver just weeks after playing on the winning team in the College World Series, and he'd been doing rehab for four years with nothing to show for it when he signed up for her trial. With a set of electrodes about the size of a shoestring French fry humming in his lower back, he spent countless hours in a standing frame or suspended over a moving treadmill while helpers placed his feet in the motion of stepping. He recovered enough motor function to stand unassisted and take supported steps under his own power. But that wasn't all.

What the questioner wants to know, since Jerry just gave a whole talk about successfully getting rats to pee, is whether Jerry has heard that Susan's patients have randomly regained control of their bladder and bowel functions. Jerry says yes, and that he thinks this is yet more evidence about what he refers to as "primitive systems." That phrase refers to basic biological functions, like getting rid of pee and poop, as well as sexual function. Jerry believes that the central nervous system really wants to keep all of that stuff working, which would account for there being no genetic instruction to those axon tracts to not grow back.

He sees Harkema's results as a sort of parallel to his own: that if you just reach a certain threshold of potential, the axons will find a way to get through they'll figure out how to rebuild the necessary connections. He goes on to say that one of his compatriots works at the Veteran's Administration at Case Western University. This man has told him stories about some patients whose diaphragms respond to their pacemakers by healing themselves, to the point that they no longer need those pacemakers. The buzzword is *plasticity*, which in the context of the brain and the body is about a serious capacity to adapt and change – to behave like *plastic*, in the sense of something very easily shaped to serve a variety of purposes.

What's really cool is that the brain figures this stuff out. All these axons that are regenerating? We're not controlling them. They're going where they may. And yet the output that we see in the animals is improving. They don't get worse, if we do things right. I'm just really optimistic, especially after hearing this talk, that the brain can figure stuff out, so maybe we don't have to be so, so perfect in our regeneration. We just gotta get some axons to go.

Amen to that.

* * *

The last two questions have to do with something Jerry mentioned earlier about *propriospinal* axon tracts being similar to the ones that control bowel and bladder in the way they responded to the triple combo therapy. Propriospinal tracts are bundles of axons that originate in the brain; their job is to interconnect various parts of the cord to one another. You'd expect that they have something to do with motor skills, and it's possible this is why some of Jerry's peeing champions also got back a little motor function. That didn't happen at all in the chronic phase. Nothing at all in terms of walking, but definitely return of bladder function.

He wants everybody to take what he's figured out and expand on it, which is what's going on now in his own lab. Add anything you can think of to that triple combination. Add locomotor training. Add gene blockers. Add salmon fibrin. Everybody, just do it.

I'm inviting everybody I see, everybody I talk to, everybody that hears about this – to add their expertise, their strategies, to try to get other axons to grow. And I'm a firm believer that once they're past the scar, they keep going. And it looks like the brain can really help fix stuff that's even malwired.

Listening to him and thinking about the past thirty-six hours, the old Chinese proverb pops into my head: *May you live in interesting times.* We do live in interesting times, and what we're witnessing is the slowly – much too slowly – forming shape of a cure.

More Stem Cells

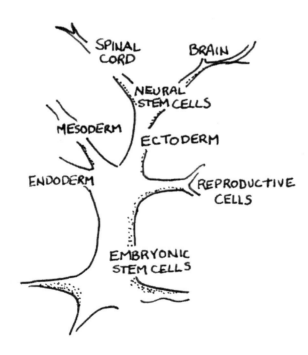

chapter Twenty

Mark Tuszynski: A Living Relay System

And Paul persisted, pretty much against my advice. And the findings that he generated have completely changed the direction of our laboratory, given the significance of the biology that he helped us to appreciate.

Mark Tuszynski

We're down to the last part of the main ballroom program; two speakers left. First is Professor Mark Tuszynski (pronounced *tuh-SHIN-ski*), who is here from the University of California San Diego, where he directs a busy lab called the Center for Neural Repair. He fits my caricature/image of a scientist: serious, slightly wild hair, energetically curious, he's like a very tall, very well educated version of Michael J Fox. His talk is about some newly published work that involves a neural stem cell strategy to repair spinal cord injury, which means it's going to have important things in common with what Aileen Anderson and Stephen Huhn told us about yesterday.

The challenge at this stage of the conference is in the struggle not to let all the information run together into one big slushy pile of I-think-that's-good-news-right. There's a line of advice in the welcome packet to Working2Walk that goes like this:

Question Authority. The scientists who have given up a few days of their time to join us want to share their knowledge. They're not here to sell snake oil; they're here to tell us what has been happening in their labs. It's everybody's job to make sure we understand what their slides mean, because that's how we arm ourselves to carry the message out into the world.

It can be very hard to do that in the moment, because you can only process so much new information at a time. I'm conscious of a certain weariness in the room, which is unfortunate: this presentation is going to be both excellent and important. Mark begins by saying he thought it was a mistake to go in the direction he's about to describe. He didn't want to do neural stem cells. A lot of people have tried and failed to get stem cells to work in spinal cord injury, and so when one of the scientists who works in his group approached him with another stem cell idea, he discouraged the guy. That scientist, whose name is Paul Lu, did the work anyway, and the results of his experiments took Mark completely by surprise and re-oriented his whole lab. He spends a few minutes naming the familiar litany of reasons why injured human spinal cords don't get

better; a lot of the people in the room could probably write this list themselves by now.

We also have a sense of what sort of approaches people are taking to go after the things on that list. Peptide injections to neutralize receptors that stick axons in phony synapse traps. Knocking down PTEN and other genes that tell axons not to grow. Bridges fortified with fibrin and growth factors to give axons a path across the cavity. ChABC to take care of proteoglycans. Cleverly engineered scaffolds that can time-release whatever is needed. Re-routing of functioning nerves to new target muscles. And, of course, cells. Neural stem cells.

The idea of putting stem cells into the injury site itself is really about forming a sort of relay system. Remember the tennis balls and strings metaphor? The tennis balls are the cell bodies, safely tucked away in the brain. The strings are tracts of axons, dangling down from those cell bodies but not able to get past the injury site to where they could connect with the nerves that lead to muscles. What if you could put some neurons inside the cord that sent out their own axons to make connections both up toward the brain and down toward the nerves? They could form a sort of relay, like runners jumping into the middle of a race.

The likeliest candidate for that kind of relay-forming cell is a neural stem cell. Aileen already gave us the image of cellular differentiation as a tree with a mighty trunk and four major branchings. The neural branching was the source for the cells she used, and – in a slightly different form – it's also the source for the work Mark's team did. One big difference between them is that she used human neural stem cells, even in her work with mice and rats. Mark's colleague Paul Lu started with rat neural stem cells. Aileen's cells were collected between sixteen and twenty weeks gestation, and his were collected in what would be the equivalent of the first trimester for a rat, at the fourteenth embryonic day. So, her experiments used second-trimester human neural cells, and his used first-trimester rat neural cells. His reasoning is that cells in the first trimester may have greater potential to survive grafting and to extend axons.

Another difference was in delivery method. Aileen's group did four tiny injections, two just above and two just below the injury site. Mark's team used a fibrin gel that was spiked with growth factors. He says that it took them a year to get a gel concocted with a combination of growth factors that would allow the cells to live and proliferate. She did her injections at nine days post injury, and he did his gel implants at two weeks. She was working all along with a biotech company eager to move the therapy to people, and he's working in the context

of a university, depending on funding from the NIH, the state of California (through CIRM and the Roman Reed Act), the Veteran's Administration, and private foundations.

Writing about these researchers many weeks after the fact, with plenty of time to go through their presentations slowly, is a very different experience than sitting in a room listening to them for two straight days, one after another. The idea of a giant, carefully coordinated Manhattan-style project begins to seem not just intriguing but – maybe – *necessary*. Aileen and Mark are just two out of who knows how many people, all working away to solve the same problem with the same basic approach and not sharing methods or strategies until after they've published their data, if they get that far. It doesn't make sense. As Os told us, sharing data is not collaborating.

Anyway. What happened when Mark's team finally got the gel right? They were working with complete transections at T3, which normally means no axons making it through the injury and certainly no motor recovery. After six weeks they sacrificed the rats and processed their spinal cords – meaning they cut them into 40-micron thick sections for viewing under microscopes. (The thickness of a single hair from a blonde newborn is about 40 microns.) The sections show in spectacular colors what happened to those neural stem cells.

The possibilities were that they would remain as stem cells, they would die off, or they would differentiate into some combination of the three cell types in the central nervous system: neurons, astrocytes, and oligodendrocytes. What actually happened was that about a quarter became neurons, a quarter became oligodendrocytes, and about fifteen percent became astrocytes. Mark doesn't say what happened to the other thirty-five percent. In Aileen's studies, the ratios were a quarter neurons and half oligodendrocytes, with very few becoming astrocytes. Why were these ratios so different? It's not clear.

What is clear is how lively those new neurons were. Their experiment produced axons in extravagant numbers, hundreds and hundreds of them. Mark: *We have 1000-fold more of these axons from these neural stem cell grafts than we had seen in any previous studies trying to regenerate the host axons themselves.* He's saying that at least in this study, it worked much, much better to build an axon relay station than to try to get the original axons – the ones alive and well but stuck at the edge of the injury site – to wake up and grow across.

The relay axons weren't bothered by all those growth-inhibiting molecules that smother and stifle mature axons. Proteoglycans? *Meh*. The relays grew up and down the cord, sometimes all the way down to the lumbar region at the base of the spine. When Paul Lu repeated the experiment with a cervical injury, the axons grew right up into the brainstem and all the way down to T8. They even grew out into the nerve roots, heading for connections with peripheral nerves. The healthy cells in the damaged rat cord responded to these axons in the normal way: existing axons formed synapses with the new ones, and existing oligodendrocytes wrapped the fresh axons in myelin, just as if they'd been waiting for something to do.

Mark shows us beautiful images of axons growing down from different areas in the brain and forming synapses with their relays – but not all areas of the brain. The ones that grew into the injury site came from the *reticulospinal* the *raphespinal*, and the *cerulospinal* areas. At the time of this writing (in April, 2013) it wasn't clear whether *corticospinal axons* grew as well. Remember Bob Yant telling us that every single scientist he surveyed told him we're never going to end paralysis until there's a functioning corticospinal tract full of working axons? That's what we hope to see.

The rats' motor recovery does point in an encouraging direction; those treated rats – with transected spinal cords – could move their hind limbs. They could bend their ankles, knees, and hips through the normal full range of motion. This isn't enough to bear weight or walk, but it's definitely more than should be possible with a transected cord injury.

Mark and Paul also tested for an electrical signal getting across that transected cord; normally with a complete transection there wouldn't be any way for a signal to jump across. The process for this test goes like this: stimulate the cord a few segments above the injury and watch for a response a few segments below it. When you do this with an intact cord and graph the results, you get an image that looks like the side of a steep, smooth hill. When you do it with an untreated transection, you see a flat plain. When Mark did it after the gel/nuclear stem cell transplants, the image was more like a jagged peak than a smooth hill – but clearly there was a signal. The relay switches were working.

Just to be sure they weren't kidding themselves, they re-cut the spinal cords just above their grafts. *Mark: We've often been fooled by things in this field. And one way to get more confidence that what we're seeing is actually valid is to do things like, after you see this recovery, we'll re-cut the spinal cord.* As you'd expect, the electrical signals went flat, and the small amount of motor recovery vanished. Then what? Having proved that something like a functional relay was doable with rat neural stem cells but not really caring all that much about rat spinal cord injuries, they moved on to trying the same thing with human neural stem cells, just as Aileen's team had done.

More transected rats, more gel implants, this time with two varieties of human neural stem cells, neither of which matched exactly the ones in Aileen's experiments. The result?

Mark: I have to tell you, to a scientist who's worked in the SCI field, to see this number of axons growing below a complete transection site, to me I still find it astonishing that this kind of growth is possible. These are many axons and they're growing in the white matter, that for decades we've studied as being inhibitory. They're not inhibited, and they're growing for long distances. These are human neural stem cells . . . we see them in these very high densities, emerging from the injury site and traveling below it . . . axons emerging from the human neural stem cells in the rat spinal cord.

They form connections . . . they show a very similar, almost identical degree of recovery, 7 on this scale, and then if we re-transect, once again, we abolish the recovery. Here's the other human stem cell line. We put this into a mid-cervical injury site – c5, a half spinal cord lesion . . . the cells live in the lesion, here they are, and you can see again this other human stem cell line is extending many of these green axons below the lesion. And they extend them up as well. So what you see here are these axons growing into the brain from the site of the implant, even as far up as the cortex of the brain.

There's really nothing quite like seeing a scientist get excited about his work.

* * *

So, where does this go? CIRM just awarded Mark's group almost $5 million to move this therapy forward, with the intent of bringing it to humans – to us – as quickly as possible. From Mark's perspective, that means testing the heck out of it in primates. He's in sharp disagreement with Aileen here, who said yesterday that she thinks large animal models are a waste of time and resources. Mark says that 95% of things that worked in animals don't work in humans, which is discouraging to hear after all this good news. Which of them is right? It's not obvious to me.

He gives an example to support his position. The gel compound that worked so well in his rats just washed away in the monkey's cords. It had to be completely re-engineered, but they couldn't have learned exactly how without trying it. The injury model they're using with the monkeys is a c7, which is the most common injury site in people. Because the work is ongoing (meaning not yet peer-reviewed and published) he only shows us a few preliminary results.

It's the same as in the rats: human neural stem cells turning into neurons, which grow axons that bloom up and down the cord. Existing axons growing out of the primate brains and down into the relay gel, forming synapses as they go. There has been a lot of interest in their methods from around the world, and

many young scientists have come to the lab to learn from Paul Lu and hopefully carry this forward quickly. Again, I can't help but think that now just might be the right time, finally, to build that globally coordinated effort.

He ends with the remark I wanted most to hear: that they're in the process now of beginning trials with chronic injury models in primates.

chapter**Twentyone**

Hans Keirstead: Something Old, Something New

> *Imagination is everything. It is the preview of life's coming attractions.*
> Albert Einstein

Our longtime friend Hans Keirstead will be the last speaker at the full conference. He's worth waiting for, as always; Hans is a force of nature. Canadian (Nova Scotia) by birth, he's obviously become very much at home here in southern California. He's telegenic, he radiates an easy intelligence, and he's good at explaining complex matters. He could have been a newscaster, but also there's something of the evangelist in him – except he preaches an intricate, neurological gospel that probably won't make it into any megachurch sermons soon. When Os Steward was first building the Reeve Irvine Research Center team back in 2000, a very young Hans Keirstead was the first scientist

he recruited. His talk today, he says, will cover a couple of old projects and a couple of new ones.

First up will be a short riff on the Geron trials. Just as Stem Cells, Inc. is the company currently running human trials based on Aileen Anderson's lab work, Geron was the company that ran human trials based on what Hans Keirstead did. The trials were shut down not because there was a problem with the science or a bad safety outcome, but because Geron ran out of money to support them.

It's hard to comprehend how monumental was the achievement of getting those trials to happen at all. First, the science itself is stunning. Hans and his team took human embryonic stem cells – the ones that form the trunk of the tree, before any branching happens – and developed them *in vitro* into oligodendrocyte precursors. They moved the cells along a development path until they had them exactly at the place on that branching tree where they wanted them, which was just at the moment when they'd abandoned all possibility of being either neurons or astrocytes. They weren't neural stem cells anymore. It's a different strategy from the one Aileen and Mark Tuzsynski used; both those scientists harvested cells that had traveled the differentiating path inside a living, gestating embryo. Hans took the cells from one of the first embryonic stem cell lines and recreated that gestational path in a lab dish. He was the first to get this unimaginably complex process working.

Then he did experiments with rats to show that his cells were functional, in the sense that they did what oligodendrocytes are supposed to do: build myelin coating around any axons in the neighborhood. The therapy idea was to take advantage of the fact that a damaged spinal cord usually has many surviving axons that cross the injury site just fine but that fail at getting messages from the brain because they've lost their insulation. When he put the human cells into rats with subacute (ten days to two weeks old) injuries, the rats got back a lot of motor function. Unfortunately, when he tried his cells with chronic injuries, there was no recovery.

If he'd stopped there it would have been a remarkable feat, just in terms of the science. He didn't. He wanted to get his cells into human beings with new injuries, and that's where every last ounce of personality, determination, and ability to look around corners was needed. The Geron trials only happened at all because Hans was so intent on meeting whatever impossibly high regulatory bars the FDA could set. He's telling us now that it was good that the bars were high. The FDA didn't know how to regulate this new thing that was neither drug nor device, and if they'd made a mistake and let an unsafe treatment happen,

> In 1999, Jesse Gelsinger was a recent high school graduate from Tucson, AZ. He had a rare genetic disorder that interfered with his health, and he volunteered for a clinical trial that involved a gene therapy. Instead of helping him, that therapy killed him within just a few days. His death led to a much more vigilant FDA.

it could have set back the whole field for a long, long time. The Geron therapy needed to do no harm.

And it didn't. When the many, many thousands of pages of supporting data had been exhaustively reviewed and the multiple holds had been lifted, five patients with new injuries were finally treated. They each got little tiny test-size doses of the cells – only a tenth of what would have been needed to see any functional benefits. All the patients were fine, and are still fine; one of them is actually here at the conference. As would be expected with so few cells, they also didn't see any benefit.

In the end, Geron made a business decision to abandon the effort. Going forward would have meant a vista of multiple millions of dollars invested with little expectation of return, given that each patient had to be followed for fifteen years, as would the next cohort, who at least would get therapeutic-sized doses and have some hope of recovery. Hans tells us that he hasn't given up on this therapy and that he's been closely involved with a couple of groups who are even now trying to resurrect it.

* * *

So that's one old project. The other update – and this is typical of Hans – is about a therapy that goes in a completely different direction. It has to do with motor neurons. Anatomy reminder. Remember the tennis balls and strings analogy? There are neuron cell bodies in the brain, and they send their axons down in bundles called *tracts* that form what's called the *white matter* of the spinal cord. The spinal cord isn't just white matter, though. It also has its own neuron cell bodies that make up the grey matter of the cord.

These grey matter neurons also send out axons, and those axons twine themselves into *spinal nerves* that shoot out of the cord and into the rest of the body. The grey matter cells deep inside the cord are called *motor neurons*. Their axons extend all the way to the fibers of our muscles, where simple chemical reactions between the tip of the axon and the muscle fiber cause the muscle to contract.

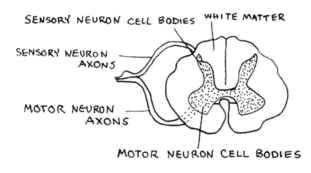

We need both. We need the cord's white matter and we need its grey matter. We need the axons coming down in tracts from the brain, and we need the motor neurons that connect with them, pick up their commands and carry them out to our muscles. So, what does all this have to do with Hans' other therapy?

There's a terrible genetic condition that happens to about one in 10,000 babies. These babies have a mutation in the DNA of their fifth chromosome; right where there should be a recipe for a protein called the Survival of Motor Neuron (SMN) protein. As you might guess, when that recipe is working, we get a nice protein that makes motor neurons live long, happy lives. When the recipe is garbled, the motor neurons die. And when that happens, babies born with this condition can't breathe. Their brains are fine, their white matter tracts are fine, but their diaphragm muscles never get the chemical jolt that a working motor neuron axon would supply.

This condition is called SMA-1, for *Spinal Muscular Atrophy #1* – the worst possible form of this condition. The second old project Hans is going to talk about is a potential treatment for SMA-1. Hans is a spinal cord injury researcher, though, right? Why is he thinking about SMA babies? Because, as he told me years ago, it would be faster to get the FDA to try a new stem-cell-based therapy on a fatal condition like SMA-1 than on (otherwise perfectly healthy) people with spinal cord injuries. The babies usually don't survive past infancy, so the urgency to find a way to help them (not to mention their suffering families) is real and pressing.

His strategy was to use human embryonic stem cells – the ones at the trunk of the tree – to make motor neurons. He'd put these healthy motor neurons into the bodies of the SMA-1 infants, and if the treatment worked those babies would be able to breathe. Having shown in the process that his motor neurons were safe and effective, he would be positioned to get permission to test them in spinal-cord-injured people whose grey matter had been damaged by injury or disease. All that work has been done – all the growing of cells, all the animal testing, all the production of documentation, along with sponsorship from a company called California Stem Cells, where Hans is the chair of the scientific advisory board. He's confident that this therapy will finally make it into human trials in 2013, both in the UK and in the USA.

Anything could happen. What I know for sure is that he's never going to give up. And, it would be a mistake to bet against him.

* * *

So what about the new projects? There are two, and both are in the early stages of development, which means that he's already done a couple of years' worth of work on them but has not published results. He won't be sharing data. The first one has to do with that molecule called PTEN. Os told us yesterday about how deleting the *PTEN* gene – and thus destroying the recipe for the PTEN molecule – allowed corticospinal axon tracts to grow. In collaboration with Os, Hans has been working on finding ways to deal with the PTEN molecule on a time-limited basis – not knocking it out forever, but just for a few hours. Why not just get rid of it? That would be a mistake, says Hans.

Hans: You can't go adding a PTEN inhibitor to people. PTEN's a tumor suppressor. It's gonna cause cancers. So our idea here was to treat cells in a dish, clean it all out, and then put those cells in you. So that we get the supercharge of the cells and the neurite outgrowth without the toxicity and detriment that PTEN inhibitors would cause in the human. And it worked.

The idea is that PTEN molecules are in our bodies for good reasons, and we wouldn't want any therapy that permanently destroyed them. Who would trade a spinal cord injury for cancer? On the other hand, those molecules are standing between brain-to-spinal-cord communication, so they have to go. In the experiments Os talked about, they delivered a knockout molecule to the part of the rat brains where corticospinal neuron cell bodies are. It worked, too, but it can't become a therapy for people until there's a way to make sure we still have our PTEN working for us once the axons have grown back.

Hans has been working with Os on how to do this. His idea was to (a) grow a neuron precursor cell from a human embryonic stem cell line, just as he's already grown other kinds of cells from those lines; (b) use a blocker molecule to delete the PTEN from those cells in a dish; (c) put the PTEN-free neuron precursors into injured lab animals; (d) wash the blocker molecules out of the transplanted cells after a short time period; (e) verify that the animals still got the benefit – namely, that the axons grew.

Easy to write those steps. Very, very hard to do. But, says Hans, "*We do confirm that we can wash out the PTEN and see a lasting effect on neurite outgrowth.*" And that's where this project stands at the moment. Stay tuned.

* * *

The other new approach sounds like it shouldn't be possible. Inside a developing embryonic spinal cord there are cells called astrocytes, and those cells play a big helper role. They're there alongside the neurons and oligodendrocytes – in fact there are five times as many astrocytes as neurons. You could think of these young astrocytes as guides or teachers of neurons. They offer structural support, and they produce molecules that show embryonic neurons which way to send their little axons.

Once the brain and spinal cord are all built, mature astrocytes play another set of roles, mostly helping neurons function well. They produce

molecules that are fuel for neurons, they help pass along the neurotransmitters needed for synapses, and they play a role in getting rid of the excess potassium that's one cause of epileptic seizures.

After an injury, these same little star-shaped cells are the builders of the prison. Remember the proteoglycans that chABC takes down? Those are a gift from post-injury astrocytes. Hans's new idea, he says, is something he first thought of about fifteen years ago. What if you could somehow turn back the clock on those cells? What if you could take a population of post-injury astrocytes and hit a rewind button that made them just like they were in the good old development days? They're the very same cells, after all.

> Cells that have been re-programmed to behave like embryonic stem cells are called *induced pluripotent stem cells*, usually abbreviated as *IPS cells*.

This pure science fiction became a lot more possible in 2007, when Shinya Yamanaka figured out how to take a human skin cell and – by adding just four genes to the DNA in its nucleus – turn it into an embryonic stem cell. Yamanaka had essentially gone all the way out to a leaf on that metaphoric development tree and forced the leaf to go back in time -- to take on all the characteristics of the trunk that produced it. It wasn't a leaf anymore. If there were a scientist who had mastered the process of taking it back up the tree, it could now become any other kind of cell in the body – or maybe just a younger and more vigorous version of its own self.

Hans has been busy becoming that scientist. He's taken astrocytes from injured spinal cords and turned some of them into cells that look and behave very much like development-stage astrocytes.

Hans: We started out with 30,000 genes and reduced it down to a number of genes that are typical of a young astrocyte and not present in an old astrocyte. And we used these things as targets to reprogram and introduce these young genes into the old astrocytes. So in effect what we're doing is pelting the genome – pelting the cell – with dozens and dozens of copies of genes that are not usually expressed in an old astrocyte; they're actually expressed in a young astrocyte. And the old astrocytes are tricked into thinking that they're young – they back-differentiate and become young. That was the goal, and I think we've managed to do it.

He goes on to show us all the reasons to believe this process has worked; it wasn't possible, obviously, to ask the astrocytes to raise their hands if they were feeling young. They could, however, make lots of comparisons. Old astrocytes don't proliferate, young ones do, and their reprogrammed ones did. Check. Same story with ROC inhibitors, laminin expression, interactions with dorsal root ganglia, and migration after scratch tests. The re-programmed cells also just looked much more like young cells than like standard old cells.

He tells us that not all the cells went back to their early state; in fact,

only pockets of them did so. On average, there were fewer effectively reprogrammed cells than ones that resisted the process. He's currently working out how to use what's known as short hairpin RNA (shRNA) to target a few specific genes that are preventing cells from going back in time.

So. All of this work so far has been happening in dishes. They're just now getting started with animal models to see what happens when a bunch of old astrocytes get fooled into thinking they're actually living in a developing central nervous system and not in one that's been damaged. Theoretically, they'd get to work and do what they do in the embryonic spinal cord: guide and support the growth of neurons. And they would stop producing proteoglycans, which means the injury site would become a much friendlier place. I can't wait to see the data from these experiments.

This presentation makes me think of a quote from an IBM information processing expert named Emerson Pugh: *If the human brain were so simple that we could understand it, we would be so simple that we couldn't.* Maybe Emerson Pugh should be introduced to Hans.

chapter**Twentytwo**

Shots on Goal

You miss 100% of the shots you don't take.
Wayne Gretzky

Mark's associate Paul Lu takes a seat next to Hans at the presenter's table; the two of them are about to take questions on the last couple of presentations. The first one comes from the audience, for Hans. Basically, the questioner wants to know if it's ever going to make business sense for some company to take up either of the new therapies he's trying to develop, or if a Geron-type funding problem is inevitable. Hans delivers a long, nuanced answer to this effect:

- All over the world, people are seeing some kind of good results when they introduce "young" cells to the injured cord in preclinical

- (i.e., animal) studies.
- The problems associated with getting those therapies to human trials are immense.
- There will be places – depending on what the therapy is – where it's cheaper and quicker to get to human trials.
- Some approaches will be very risky, and that means tough controls, and that means money.
- It's awesome to have a bunch of scientists making shots on goal; just like in hockey and soccer, it's the only way to score.
- The challenge then becomes that scientists are drawn into a high-octane strategy game that involves guessing at risk and cost along with potential benefit.
- Scientists need to be able to do their work without worrying about that game, because that's how you get them to produce a lot of shots on goal. That said, they can't just ignore the question of how their work could make it into the market.

Now someone asks Paul if he doesn't think that they got *too much* axon growth from their transplanted neural stem cells. Surely all those axons aren't going to make functional synapses, so does Paul think they'll adjust the dose going forward? Paul replies that the questioner is right. In the beginning they transplanted tons of their neural cells into the transected cord – enough to completely fill up the cavity. Because almost all human spinal cord injuries are contusions and not transections, the eventual dose would be much smaller.

He goes on to talk about how during development, axon-growth is very highly organized and guided. The neural stem cells they're putting into the lab rats are great at growing axons, but will the rat bodies execute the guidance program the way they did during gestation? Or will the scientists have to figure out how to control the growth themselves? As they say, further study is needed. Another question they're trying to answer is what happens to all the axons over time – which means doing the grafts and then waiting a year to sacrifice the animals and use their staining techniques to find out.

Someone else wants to know if the ratios of cell types that grew out of Paul's transplants are the same as the ratios that naturally appear in fetal development. Paul says that he thinks they're approximately the same, and that the only thing they do to their neural cells before they transplant them is to add some growth factors. Because the growth factors are proteins that naturally wash out in a day or two, he doesn't think they would have any effect on what proportions of the stem cells become which of the three possible cell types.

Speaking of growth factors, there's a question. Can we consider that part of what's needed for a future combination therapy as *done*? Can we scratch it off the master list of things that still need work? Paul doesn't quite answer this question, apparently because it's based on a bit of a misconception. He tells a story about the early efforts of the Miami Project scientists to build grafts with Schwann cells, and then shifts to making a quick list of all the things that might fall into the category of growth factors.

Early stage neural stem cells need some kinds of growth factors just to stay alive. They also need what he calls *proliferation factors*, which are molecules that stimulate them to divide a few times and fill up the cavity before they start differentiating into one of the three central nervous system types. And, there will have to be some molecules that cause new blood vessels to form, because in a graft you need extra nutritional support, and it's the blood's job to deliver it. He says that one of the things they're doing in the lab right now is testing individual factors and combinations, in search of just the right mix.

So, no. We can't cross that one off the list yet.

The last question is the one we all came here with. *When are we going to see any of this tested in humans? How long before any of it gets to clinical trials?* Hans replies that he talked about four things: the Geron therapy has already been tested in humans and he believes will continue to be tested; the SMA-1 therapy will get to human babies during the first half of 2013; the other two are pre-clinical and will take "a little while" before they get to people. He points out that the more he does this, the faster it goes. The SMA-1 therapy took $1/8^{th}$ of the time and $1/40^{th}$ of the money that the Geron trial required.

Paul says that thanks to CIRM and the Roman Reed funds, they have money to keep going and get the large animal studies done, and that's going to take a three or four years. It's a sobering end to the day.

* * *

And that's it. The end of presentations from scientists, every one of whom is obsessively trying to cure people who live with paralysis. We'll stagger and roll off to our rooms and to the hotel bar now. We'll get together in restaurants for takeout food and talk about what we heard and what we didn't hear; it's good, it's encouraging, it's confusing, it's *so not enough*. We'll strategize about how to make it all go faster. Tomorrow there will be a visit to the Reeve

Irvine labs and workshops, and after that a lot of airplane rides to homes all over the world. And then we'll start making plans to be together again in Boston, at Working2Walk 2013.

chapter**Twentythree**

Please Come to Boston

There is no real ending. It's just the place where you stop the story.
Frank Herbert

It should be obvious by now that there's still work to be done. The 2013 Working2Walk Science and Advocacy Symposium has been in the planning stages since the moment we all straggled onto planes leaving Irvine. This year we'll be hanging out at the Boston Convention and Exhibit Center on Friday and Saturday, September 27th and 28th. Friends will be made. Drinks will be had. Lives will be changed. The agenda is still under construction, but confirmed speakers include the usual array of advocacy and research stars. Among many others, I'm personally looking forward to news from Neuralstem

CEO Richard Garr, whose company was given the FDA go-ahead in January 2013 to test a cell-based therapy on people with chronic, complete spinal cord injuries. Have they done their first patients? What results can be shared?

Richard Garr was with us at the October, 2011 conference in DC. The blog I wrote that day describes him telling us about Neuralstem's efforts to get FDA permission to test their cells in people with ALS, which – he hoped – would lead to trials with us. He talked about the government's (understandable, frustrating and appreciated) obsession with safety. He said that Neuralstem was also looking at doing trials in India and China – also on chronics – but that the USA work would have to come first. This therapy, for those who missed hearing about it the first time, is similar to what Aileen Anderson and Stephen Huhn are testing on spinal-cord-injured people right now in Zurich. It's similar to what Mark Tuzsyinski is testing right now on monkeys in California. I can't tell you how strange it feels to write this, after all the years of promises and hopes that never made it to trials on chronics.

I'm going to be there. And I'm looking forward to the day when we don't need to do this conference anymore – the day when people are walking, and running, and making love, and dancing. We have a role in bringing that day about; nobody's going to do our part for us. Luckily, though, it's fun to be an advocate. Luckily, as Betheny put it, *"Spinal cord injury is a rough way to meet people, but I've met some keepers!"* Amen, sister.

Sources and Resources

Unite2FightParalysis (www.u2fp.org)

This is the group that organizes Working2Walk every year; their website is lively, informative, and easy to use. Go early, go often.

Working2Walk Videos (http://vimeo.com/u2fp/videos) and (http://www.youtube.com/user/unite2fightparalysis)

All of the videos on which this book is based (along with many others) are available online at u2fp's youtube and vimeo channels. The easiest way to get to them is by going to the u2fp website and clicking on the *Vimeo* or *Youtube* options under the *Connect* list that sits at the bottom right of every page. **These presentations (and the papers on which they were based) were my primary source.** I highly recommend that readers not be satisfied with my descriptions of them, which are necessarily filtered and constrained. All the scientists who presented papers were invited to review the chapters that dealt with their material, and most of them generously gave yet more of their precious time to help make this book accurate and useful. (Any errors or omissions remaining are of course mine and only mine.)

Live Blogs of Working2Walk, 2007 - 2012

http://working2walk2012.wordpress.com/ (2012 Irvine CA)
http://sci.rutgers.edu/forum/showthread.php?t=167114 (2011 Washington DC)
http://sci.rutgers.edu/forum/showthread.php?t=142497 (2010 Phoenix AZ)
http://working2walk09.blogspot.com/2009/08/anne-phipps-co-founder-sci-step.html (2009 Chicago IL)
http://working2walk.wordpress.com/ (2008 Washington DC)
http://sci.rutgers.edu/forum/showthread.php?p=644352#post644352 (2007 Washington DC)

The live blogs are my attempts over the years to capture the symposiums as they unfolded. I started writing them in 2007 with people who couldn't travel in mind. We were too poor to make the trip ourselves to that first rally in 2005, and I knew how frustrating it was to know something so important was happening and get no news of it. We did make it DC in 2006; I took notes and tried to write them up later, but that was unsatisfying. I got a little better at the blogging thing over time, I guess. The 2012 blog has had thousands of page views from more than fifty countries.

Working2Walk 2012 Presenters

Aimetti, Alex http://www.epernicus.com/aaa5
Anderson, Aileen http://www.anatomy.uci.edu/anderson.html
Bellamkonda, Ravi http://www.bme.gatech.edu/facultystaff/faculty_record.php?id=59
Blackmore, Murray http://www.jbixbylab.com/Pages/murray.html
Brown, Justin http://surgery.ucsd.edu/faculty/Pages/justin-brown.aspx
Havton, Leif http://www.anatomy.uci.edu/havton.html
Huhn, Stephen http://www.stemcellsinc.com/about-us/people/leadership-team.htm
Keirstead, Hans http://www.anatomy.uci.edu/keirstead.html
Silver, Jerry http://neurosciences.casc.edu/faculty/silver/index
Steward, Os http://www.anatomy.uci.edu/steward.html
Tuszynski, Mark http://tuszynskilab.ucsd.edu/tuszynski.php

The web pages referenced here are a starting place for those who want to know more about the achievements of the 2012 presenters. Every academic paper these people author represents years of effort, decades of hard-won knowledge and hundreds of thousands (in some cases millions) of dollars.

CareCureCommunity (www.sci.rutgers.edu)

This is the online message board that's been home to so many of us for so many years. Thanks to Wise Young (who himself has published dozens of articles and almost 40,000 posts there), a lot of people who might otherwise experience life isolated are instead valued members of a functioning virtual community.

The original thread suggesting a rally to honor Christopher Reeve in January 2005 is here: http://sci.rutgers.edu/forum/showthread.php?t=20918

Reeve Irvine Research Center http://www.reeve.uci.edu/

The scientists at RIRC not only spent hours and hours explaining their work to us, they invited us to come to their labs and spend a morning asking questions. The final event of Working2Walk2012 was a rare chance to be among not just the people who run labs but also the people who work in them. We owe the grad students who care for animals much, because without them there would be no data and no progress toward a cure. I highly recommend the RIRC newsletter; it's clear, engaging, and designed for the non-scientists among us. RIRC Newsletter Archives: http://www.reeve.uci.edu/news.html

Langer Lab http://web.mit.edu/langerlab/

This is the home page of one of the most prolific chemical engineering labs in the world. Their list of publications begins in 1974 and just keeps going;

these are good people to have on our side. There's a fascinating profile of Bob Langer himself available on this New York Times page: http://www.nytimes.com/2012/11/25/business/mit-lab-hatches-ideas-and-companies-by-the-dozens.html?emc=eta1&_r=2& (Or you could just type *Bob Langer New York Times* into google. The article will be the first thing that comes up.)

Stem Cells, Inc. http://www.stemcellsinc.com/

This is the California-based company that's running a clinical trial on people with chronic injuries in Zurich, Switzerland. You can read about the trial here: http://www.stemcellsinc.com/Therapeutic-Programs/Clinical-Trials.htm and you can enroll to be part of it here: http://www.stemcellsinc.com/Therapeutic-Programs/Clinical-Trial-Sites.htm

InVivo Therapeutics http://www.invivotherapeutics.com/

This is the company that Working2Walk presenter Alex Aimetti works for. As you can read on their website, in April 2013 they were given FDA permission to begin testing one of the therapies Alex described to us. Between the time of the conference and this writing (May 2013), InVivo stock has almost doubled in value. May they live long and prosper. Don't miss their published papers, available in full at http://www.invivotherapeutics.com/our-research/

Christopher and Dana Reeve Foundation (CDRF)

This one needs no explanation. The home page is here: http://www.christopherreeve.org It's a one-stop-shop for people looking for information, company, news, video, blogs, statistics – all available in eight different languages, all constantly updated. Our collective debt to Chris and Dana goes on. One of my favorite things at the Reeve site is my friend **Knowledge Manager Sam Maddox's blog**. Go here (http://www.spinalcordinjury-paralysis.org/maddogz) for lively, current, up-to-date information about the research being funded by the Reeve Foundation.

California Institute for Regenerative Medicine (CIRM) http://www.cirm.ca.gov/

This website is like a Disneyland of information related to progress in cell-based therapies, not just for spinal cord repair, but for all kinds of conditions and diseases that are candidates for this kind of strategy. You can track the money spent on spinal cord injury on this page: http://www.cirm.ca.gov/grants?field_public_web_disease_focus_tid%5B%5D=931 but I'd encourage anybody reading this to spend some time wandering around in this site. Progress for any of us is progress for all of us.

Featured SCI Advocate Websites

Spinal Cord Injury Sucks http://www.scisucks.org/ (Geoff Kent)

Roman Reed Foundation http://romanreedfoundation.com/ (Roman Reed)

Stem Cells and Atom Bombs http://stemcellsandatombombs.blogspot.com/ (Dennis Tesolat)

Stem Cell Battles http://www.stemcellbattles.com/ (Don Reed)

Cure Medical http://www.curemedical.com/about_founder.html (Bob Yant)

International Spinal Research Trust http://www.spinal-research.org/ (Mark Bacon, Director of Research)

Get Up, Stand Up for the Cure http://gusu4cure.org/author/matthew/ (Matt Rodreick)

Research for Cure http://www.researchforcure.org/www.researchforcure.org/RESEARCH_FOR_CURE.html (Karen Miner)

Books

I'm not formally trained in neuroscience or anatomy, but I do have an undergraduate degree in math and a masters in learning theory – and I can read. The first four books on this list are the texts I depended on to help me make sense of the work presented in Irvine; the rest are included because they introduce subjects of interest to those of us trying to learn how science works and how it can help us. All of them are intended for general audiences; the writing is uniformly excellent. You won't be surprised, maybe, to notice that I'm fond of biographies of scientists.

Essentials of Anatomy and Physiology, (2005) Edited by Seeley, Stephens, and Tate

In Search of the Lost Cord: Solving the Mystery of Spinal Cord Regeneration, (2001) Luba Vikhanski

Genome: The Autobiography of a Species in 23 Chapters, (1999) Matt Ridley

The Immortal Cell, (2003) Michael D West

Time, Love, Memory: A Great Biologist and His Quest for the Origins of Behavior, (2000) Jonathan Weiner

The Demon Under the Microscope: From Battlefield Hospitals to Nazi Labs, One Doctor's Heroic Search for the World's First Miracle Drug, (2007) Thomas Hager

In Praise of Imperfection: My Life and Work, (1989) Rita Levi-Montalcini

Minds Behind the Brain: A History of the Pioneers and Their Discoveries, (2004) Stanley Finger

The Language of Life, (2010) Francis S Collins

The Emperor of All Maladies: A Biography of Cancer, (2011) Siddhartha Mukherjee

A Life Decoded: My Genome: My Life, (2008) J Craig Venter

The Structure of Scientific Revolutions, (1962) Thomas S Kuhn

His Brother's Keeper, (2005) Jonathan Weiner

Acknowledgments

On the day I dreamed this project up, I called Donna Sullivan and Marilyn Smith to ask if it would be okay with them for me to write a book about the conference that consumes most of their waking hours. I was on my cellphone, walking through my neighborhood. Uphill. I was out of breath. I didn't do a great job of describing the idea. Still, they said, *"Go for it."* They trusted me to get it mostly right, and I owe them for that. They, along with fellow Unite2FightParalysis board member Chris Powell, answered questions, provided feedback, dug up documents, and generally cheered me on. I owe them for that, too.

Many scientists, in the process of giving me feedback about the technical details of their presentations, were kind enough to make comments about the writing and the project itself. I can't describe how encouraging those remarks were, especially on the days it seemed I was never going to find the right words to make their work understandable. Thanks, guys.

For nearly twenty years I've had the good luck to be part of a group of women writers. This is the place to thank them for being my friends, and especially for listening so carefully to certain chapters of this book. Judy Bentley, Janine Brodine, Susan Starbuck, Terri Miller, Christine Castigliano, Stephanie Guerra, Jennifer Bradbury, and Jackie Levin: you're collectively and individually amazing. Thank you.

The first person I thought of when I realized I was going to need help to turn this prose into something attractive and readable was my friend, Lynn Kirkpatrick. I knew her to be a good designer who was easy to work with; I didn't expect her to also turn out to be the most patient woman in the Pacific Northwest. Thanks, Lynn, for putting up with a zillion edits and for thinking up so many ways to make this a better book. I owe you.

Of course, this book only exists at all because of the generous support of my Kickstarter backers, all of whom are named up front. Thank you again and again.

Bruce Hanson – partner in all things, man who never fails me – is both the reason I'm on this ride and the one who makes it possible to live it out with so much joy. He's generous, silly, and talented. Thanks, dear, for thirty years of great times, two awesome daughters, a wonderful extended family, and the beautiful covers on this book. I love you madly.